335
Topics in Current Chemistry

For further volumes:
http://www.springer.com/series/128

Aims and Scope

The series Topics in Current Chemistry presents critical reviews of the present and future trends in modern chemical research. The scope of coverage includes all areas of chemical science including the interfaces with related disciplines such as biology, medicine and materials science.

The goal of each thematic volume is to give the non-specialist reader, whether at the university or in industry, a comprehensive overview of an area where new insights are emerging that are of interest to larger scientific audience.

Thus each review within the volume critically surveys one aspect of that topic and places it within the context of the volume as a whole. The most significant developments of the last 5 to 10 years should be presented. A description of the laboratory procedures involved is often useful to the reader. The coverage should not be exhaustive in data, but should rather be conceptual, concentrating on the methodological thinking that will allow the non-specialist reader to understand the information presented.

Discussion of possible future research directions in the area is welcome.

Review articles for the individual volumes are invited by the volume editors.

Readership: research chemists at universities or in industry, graduate students.

Henrike Heise • Stephen Matthews
Editors

Modern NMR Methodology

With contributions by

S. Appelt · B. Blümich · M. Etzkorn · S. Glöggler · U.L. Günther ·
H. Heise · G. Kervern · Ē. Kupče · S. Matthews · H. Müller ·
G. Pintacuda · Y. Xu

Springer

Editors
Henrike Heise
Heinrich-Heine-Universität Düsseldorf
ICS-6 / Strukturbiologie und Biophysik
Jülich
Germany

Stephen Matthews
Imperial College London
Division of Molecular Biosciences
Biochemistry Building
South Kensington
London
United Kingdom

ISSN 0340-1022 ISSN 1436-5049 (electronic)
ISBN 978-3-642-37990-1 ISBN 978-3-642-37991-8 (eBook)
DOI 10.1007/978-3-642-37991-8
Springer Heidelberg New York Dordrecht London

Library of Congress Control Number: 2013939356

Printed on acid-free paper

Springer is part of Springer Science+Business Media (www.springer.com)

Preface

Nuclear magnetic resonance (NMR) spectroscopy is an extremely versatile and powerful analytical tool, which is indispensable in many research fields within chemistry, physics, biology, and medicine. Applications of this technique range from routine chemical analysis to materials science and from biological structure determination to biomedical imaging. Although the technique itself is almost 70 years old, it is far from being fully exploited, as novel developments in hardware technology, pulse sequences, and hyperpolarization techniques have expanded its applicability to ever more complex systems and demanding questions. In this volume, we give the reader a flavor of the large and multifaceted range of applications of NMR spectroscopy by highlighting the recent advances in a variety of areas.

While improvements in superconductor technology have facilitated high-resolution NMR spectroscopy at ultra-high magnetic fields with B_0 strengths up to 23.5 T (corresponding to a ^{1}H Larmor frequency of 1 GHz), improvements at the other end of the scale, i.e., NMR spectroscopy at ultra-low magnetic fields, also offers intriguing opportunities. NMR spectroscopy at the Earth's magnetic field is an extremely attractive option for field applications, as no large superconducting magnets are required. Furthermore, at ultra-low magnetic fields the scalar coupling becomes the major interaction which leads to a novel type of NMR spectroscopy. In the chapter "NMR Spectroscopy for Chemical Analysis at Low Magnetic Fields," the underlying principles, recent applications, and state-of-the-art hyperpolarization techniques used for pre-polarization of nuclear spins are reviewed.

A major limitation for NMR spectroscopy is the intrinsically low sensitivity due to the rather unfavorable Boltzmann distribution for nuclear spins at thermal equilibrium. Thus, considerable effort in magnetic resonance spectroscopy is made towards sensitivity enhancement by hyperpolarization techniques, such as optical polarization, para-hydrogen-induced polarization enhancement, and dynamic nuclear polarization (DNP), a method which exploits the magnetization of unpaired electrons in stable radicals or transition metals to enhance nuclear polarization beyond the Boltzmann limit. In the chapter "Dynamic Nuclear Hyperpolarization in Liquids," the fundamental theory for different polarization transfer mechanisms in DNP is explained, and the experimental background for DNP applications is described.

Structure determination of soluble proteins of moderate size (up to 30 kDa) by multinuclear multidimensional NMR spectroscopy is now fairly standard with well-established protocols. In the chapter "NMR with Multiple Receivers," an acquisition scheme employing multiple receivers is described which allows for faster structure determination in small molecules. Further, recently employed fast acquisition schemes such as Hadamard spectroscopy, projection-reconstruction techniques, and reduced dimensionality experiments are explained.

The most severe limitation towards structure determination by solution NMR spectroscopy of proteins larger than 50 kDa is the line-broadening due to restricted molecular tumbling. Transverse-relaxation optimized spectroscopy (TROSY), originally developed by Wüthrich for amide-protons, relies on the selective detection of the spin state for which dipolar coupling and chemical shift anisotropy relaxation mechanisms compensate each other, thus leading to reduced line-widths. In the chapter "TROSY NMR Spectroscopy of Large Soluble Proteins," the application of TROSY methodology to extensively deuterated proteins with selective protonation in methyl groups is reviewed.

For even larger proteins, protein aggregates, protein complexes, or proteins embedded in a lipid bilayer, where the molecular tumbling is further reduced, solid-state NMR spectroscopy may become a viable alternative for structure elucidation or even high-resolution structure determination. In the chapter "Solid-State NMR Spectroscopy of Proteins", basic principles of biological solid-state NMR spectroscopy as well as fundamental techniques for isotope labeling, sample preparation, and some selected applications are reviewed. In addition, recent developments in polarization enhancement by DNP for solid-state NMR spectroscopy are outlined.

Paramagnetic centers in molecules are often considered an inconvenient obstacle towards characterization by NMR spectroscopy. However, in the past decade, great effort has been made towards the exploitation of paramagnetic centers in small organometallic compounds, stable radicals, or even proteins with paramagnetic centers for elucidation of chemical or structural properties. In the chapter "Paramagnetic Solid-State Magic-Angle Spinning NMR Spectroscopy," the application of NMR spectroscopy to paramagnetic solids is reviewed: The theory of major interactions between electrons and nuclei such as the hyperfine shift, the pseudocontact shift, and paramagnetic relaxation enhancement is explained in detail. Experimental details are described and some illustrative recent examples are given.

The dynamics of nuclear spins in magnetic fields and their manipulation are the underlying common theme in many new developments in NMR. We hope that the reader will enjoy the range of aspects covered in our book. Furthermore, we would like to thank all the contributing authors for their valuable contributions.

London — Stephen Matthews
Düsseldorf — Henrike Heise

Contents

Top Curr Chem (2013) 335: 1–22
DOI: 10.1007/128_2011_304

Published online: 14 December 2011

NMR Spectroscopy for Chemical Analysis at Low Magnetic Fields

Stefan Glöggler, Bernhard Blümich, and Stephan Appelt

Abstract This chapter addresses the limits of low-field NMR spectroscopy for chemical analysis and will answer the question of whether high-resolution NMR spectroscopy for chemical analysis of solutions can be achieved with magnetic fields much lower than 0.1 T without losing the chemical information which at high field is derived from the chemical shift and the indirect spin–spin or J-coupling. The focus is on two major issues. First, the thermal spin population differences given by the Boltzmann distribution are small at low field and so is the signal-to-noise-ratio when starting measurements from thermal equilibrium. Second, the possibility of identifying chemical groups is explored at low magnetic fields where the chemical shift can usually no longer be resolved.

Keywords Hyper-polarization · Low-field NMR · NMR spectroscopy · Strong coupling

S. Glöggler
Institute for Technical Chemistry and Macromolecular Chemistry, RWTH Aachen, 52056 Aachen, Germany

II. Institute of Physics, RWTH Aachen University, 52056 Aachen, Germany

B. Blümich (✉)
Institute for Technical Chemistry and Macromolecular Chemistry, RWTH Aachen, 52056 Aachen, Germany
e-mail: bluemich@mc.rwth-aachen.de

S. Appelt
Institute for Technical Chemistry and Macromolecular Chemistry, RWTH Aachen, 52056 Aachen, Germany

Central Institute for Electronics, Research Center Jülich, 52425 Jülich, Germany

Contents

Abbreviations

ALTADENA	Adiabatic longitudinal transport after dissociation engenders net alignment
DNP	Dynamic nuclear polarization
PASADENA	Parahydrogen and synthesis allow dramatically enhanced nuclear alignment
PHIP	Para-hydrogen induced polarization
SABRE	Signal amplification by reversible exchange
SEOP	Spin exchange optical pumping
SPINOE	Spin polarization induced nuclear Overhauser effect
SQUID	Superconducting quantum interference device

1 Introduction

Measuring NMR signals at low magnetic field is not new but doing spectroscopy at low magnetic fields is a new and exciting field [1–8]. Starting with Packard and Varian in 1954, signals were detected in the early days of NMR in the earth's magnetic field despite low signal-to-noise ratio [9]. One motivation for low-field NMR is in the area of geomagnetic sensing with earth field magnetometers. For example, one of the earliest projects of the Nobel laureate Sir Peter Mansfield was to develop an earth's field NMR spectrometer in 1959. Varian proposed in 1968 to use earth's field NMR for well logging in oil exploration, which nowadays is a employed commercially albeit at higher field employing permanent magnets. Benoit [10] and Béné [11] showed first spectra with the hetero-nuclear *J*-coupling of $^{1}H^{31}P$- and ^{1}H–^{14}N-groups resolved in the earth's magnetic field. In addition to NMR signal detection with coils, other approaches for low-field NMR have been developed in recent years for spectroscopy and imaging. Among these are NMR detection by SQUID technology, which has widely been explored recently for spectroscopy and imaging [3, 12, 13], atomic magnetometers, for NMR detection at ultra-low and zero magnetic field [8, 14, 15], and diamonds with nitrogen vacancy centers [16, 17].

This chapter addresses the question of whether high-resolution NMR spectroscopy for chemical analysis of solutions can be achieved with magnetic fields much lower than 0.1 T without losing the chemical information which at high field is derived from the chemical shift and the indirect spin–spin or *J*-coupling. The focus is on two major issues. First, the thermal spin population differences given by the Boltzmann distribution are small at low field and so is the signal-to-noise ratio when starting measurements from thermal equilibrium. Second, the possibility of identifying chemical groups is explored at low magnetic fields where the chemical shift can usually no longer be resolved.

2 Signal Enhancement

If we consider a sample with one million proton spins at room temperature in a setup for NMR spectroscopy at a magnetic field of 1 T, then the spin population difference derived for protons from the Boltzmann law amounts to about three spins contributing to the net magnetization M of the NMR signal. In the earth's magnetic field at 4.8×10^{-5} T, the magnetization arises from only one in ten billion spins.

To overcome the loss in signal-to-noise-ratio in low-field NMR there are basically two approaches: either one can make the best out of thermally polarized samples, or one can hyper-polarize the nuclear spins to values far above those achieved by thermal polarization. When working with thermal polarization at low magnetic field, one can either average the signal until a sufficiently high signal-to-noise ratio is achieved, or one can increase the sample volume. But averaging is time consuming, and time is often limited. Increasing the sample volume is effective when enough sample is available, for example in well logging NMR [18, 19], but it is not an option when dealing with small and expensive samples. If neither time nor sample volume can be extended, the longitudinal nuclear magnetization can be increased by pre-polarizing the sample, e.g., in a large magnetic field, before detecting transverse magnetization at lower field. As the pre-polarization field does not need to be homogeneous, non-thermodynamic equilibrium magnetization can be generated either with the help of an electromagnet that can rapidly be switched between different values, a procedure common in field-cycling NMR [20–22], or by transporting the sample between different field strengths [23, 24]. When the longitudinal relaxation times of liquids are long enough, the sample may be shuttled manually, e.g., between a Halbach magnet for pre-polarization and an earth's-field spectrometer for detection.

These approaches of field cycling still make use of thermal polarization at higher field. Alternatively, a spin polarization higher than that reached in thermodynamic equilibrium can be generated by hyper-polarization. Among the hyper-polarization techniques for low-field NMR are Dynamic Nuclear Polarization (DNP) [25–27], Spin Exchange Optical Pumping (SEOP) [28, 29], polarization transfer due to the Spin Polarization Induced Nuclear Overhauser Effect (SPINOE) [30, 31], and Parahydrogen Induced Polarization (PHIP) [32–41]. All of these techniques have been successfully applied in low-field NMR. In the following applications of SEOP, SPINOE, and PHIP are discussed in more detail, as well as signal detection after pre-polarizing the sample.

2.1 Experimental Setups for Signal Enhancement

Figure 1 depicts the experimental setups of the polarization techniques discussed in the following: (1) shows how the sample is pre-polarized in the magnetic field $B_p = 2$ T of a Halbach magnet before it is manually transported in about 2 s to the spectrometer, where the NMR signal is acquired. (2) outlines the production of hyperpolarized xenon gas by SEOP with an Rb–Xe-flow-polarizer and the subsequent polarization transfer by SPINOE to a sample inside the NMR spectrometer. For hyper-polarization with para-hydrogen (3), the thermal equilibrium mixture of ortho- and para-hydrogen gas needs to be enriched in its para-hydrogen content. This is achieved by cooling hydrogen gas down to temperatures of 77 K or lower and bringing it into contact with an iron oxide or active charcoal catalyst. The resultant para-hydrogen enriched gas is then passed to the sample.

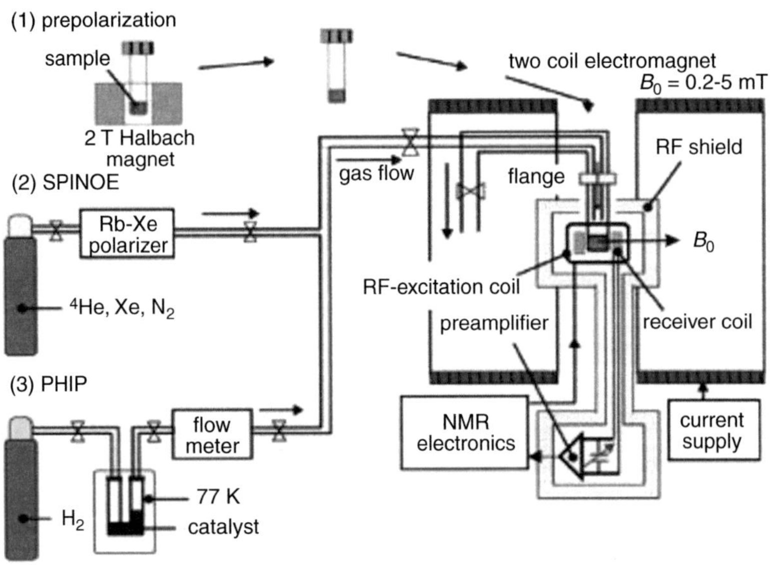

Fig. 1 Experimental scheme for polarizing samples and detecting transverse magnetization at low magnetic fields with three polarization devices: (*1*) 2 T Halbach magnet, (*2*) Rb–Xe-flow polarizer, and (*3*) gas-flow equipment to produce para-hydrogen enriched gas [41]

2.2 Prepolarizing Samples in High Magnetic Fields

Recently it was shown that highly resolved NMR spectra can be measured in magnetic fields as low as 10^{-3} T [42]. At this field strength, the Boltzmann law specifies the thermal equilibrium magnetization of one billion proton spins to arise from three nuclear spins. This low value prohibits signal detection in single-scan experiments. By pre-polarizing the same sample in a 2 T Halbach magnet the net magnetization increases by a factor of 2,000. After transporting the magnetized sample from the Halbach magnet to the spectrometer within 2 s a sufficient amount of polarization remains to acquire high-resolution NMR spectra in single-scan experiments. As an example, Fig. 2 depicts the ^{1}H free induction decay (FID) and the corresponding spectrum of 0.2 cm^3 acetic acid acquired at $\nu = 41.7$ kHz. It is remarkable that chemical shift is still resolved at $B = 9.8 \times 10^{-4}$ T.

2.3 Optical Pumping: SEOP and SPINOE

Spin-exchange optical pumping (SEOP) is a way of hyper-polarizing noble gases such as helium, neon, krypton, and xenon by transferring the nearly perfect spin order of circularly polarized photons to the electrons of alkaline atoms followed by a subsequent transfer of electron spin angular momentum to the nuclei of the noble gas. Among these gases ^{129}Xe is the most interesting one, as the atoms have a nuclear spin 1/2, and the chemical shift range is about 10,000 ppm. Moreover, the

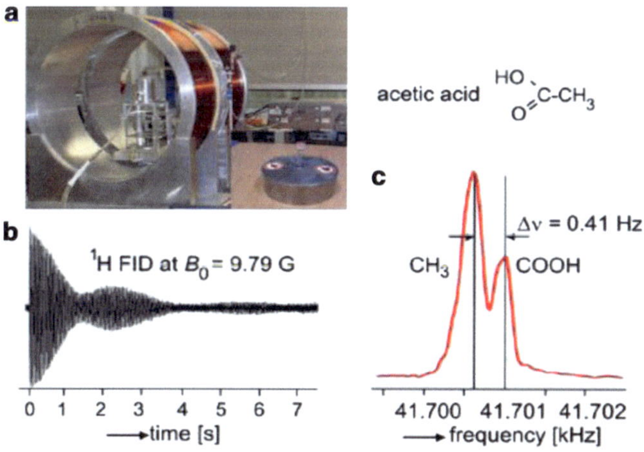

Fig. 2 NMR spectroscopy with an electromagnet at low fields. (**a**) Electromagnet for measuring NMR spectra in the milli-tesla regime and a 2 T Halbach magnet for prepolarization. (**b**) Single-scan ^{1}H FID of 0.2 cm^3 acetic acid at $B = 9.8 \times 10^{-4}$ T. (**c**) Corresponding NMR spectrum. The two lines are separated by 0.41 Hz, which corresponds to 9.8 ppm chemical shift difference between the methyl and carboxylic acid groups [42]

^{129}Xe chemical shift has proven to be very sensitive to the environment of organic liquids and bio-molecules, so that it also promises to be useful in applications at low fields. For example, it has been shown [43] that light-induced, free-radical polymerization reactions of methyl methacrylate (MMA) can be monitored in real-time by ^{129}Xe spectroscopy at $B = 3.5 \times 10^{-3}$ T ($\nu_{Xe} = 41.7$ kHz) if hyper-polarized xenon is supplied to the polymer solution. As the polymerization advances the polymer becomes stiffer, which increasingly hinders the penetration of xenon into the polymer solution, so that the signal intensity decreases. Furthermore, with increasing conversion, the solution becomes more viscous, and the chain mobility decreases. Therefore the net xenon–proton dipolar coupling is enhanced, which broadens the line. Also the chemical shift of the dissolved xenon increases with the reaction time because the xenon chemical shift is proportional to the chain length of the polymer (Fig. 3a).

Due to the large chemical-shift range, the sensitivity to organic liquids and the long relaxation times of ^{129}Xe, xenon spectroscopy can even be used for chemical analysis of sample volumes in the cubic centimeter regime in the earth's magnetic field. For example, xenon dissolved in ethanol at $B = 4.8 \times 10^{-5}$ T has a Larmor frequency that differs by about 0.016 Hz from the Larmor frequency of xenon dissolved in toluene (Fig. 3b) [4]. An outstanding observation is that the accuracy of the experimentally determined chemical-shift differences is very high with uncertainties of less than 1 ppm, although the earth's magnetic field fluctuates during the acquisition time within time frames of just 1 min. In fact, the chemical shift fluctuations measured in the earth's field are within 0.02 Hz h^{-1} comparable to those measured in a standard high-field superconducting magnet.

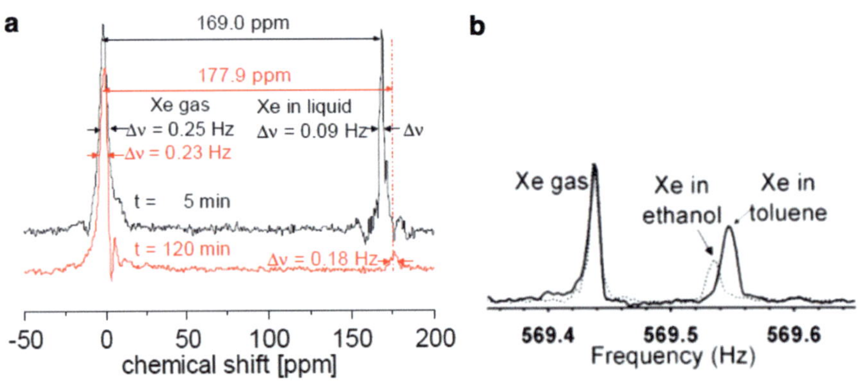

Fig. 3 (**a**) Spectrum of ^{129}Xe dissolved in a solution of methyl methacrylate during a light-induced polymerization at $B = 3.5 \times 10^{-3}$ T. As the polymerization advances the chemical shift difference between the free xenon gas and the gas atoms in solution increases together with the line width while the peak intensity decreases [43]. (**b**) Spectrum of ^{129}Xe dissolved in ethanol and toluene in the earth's magnetic field ($B = 4.8 \times 10^{-5}$ T). Even at such low magnetic fields it is possible to distinguish between the two solvents ethanol and toluene based on the ^{129}Xe chemical shift [4]

Hyperpolarized noble gas atoms, like xenon, can interact with protons and other nuclei via dipolar cross-relaxation, a mechanism which is supported by stochastic processes of motion. By means of this interaction, hyperpolarized xenon can transfer its polarization to protons and other nuclei of interest. An efficient way to make use of this SPINOE mechanism is to freeze hyperpolarized xenon onto the surface of the sample. During the defrosting process large amounts of xenon can penetrate into the liquid sample, resulting in a large SPINOE enhancement.

A further advantage of using SPINOE hyper-polarization at low magnetic field derives from the fact that the enhancement factor is inversely proportional to the field strength. Thus, large signal enhancement is achieved when experiments are conducted at low magnetic fields. A comparison between the thermally polarized proton signal of 0.2 cm^3 toluene at $B = 9.8 \times 10^{-4}$ T after 100 scans and that obtained from a single-scan experiment with Xe–^{1}H SPINOE using hyper-polarized xenon ice on top of toluene is documented in Fig. 4 [31]. The ^{1}H signal in the SPINOE experiment is larger by a factor of 1,000 than that generated by the thermal equilibrium magnetization. Although the process of SPINOE hyper-polarization is a bit complicated in terms of the setup necessary for optical pumping and sample preparation, in particular, if the sample needs to be frozen and deoxygenated, it remains a powerful approach for hyper-polarizing small samples and to detect them at very low magnetic fields.

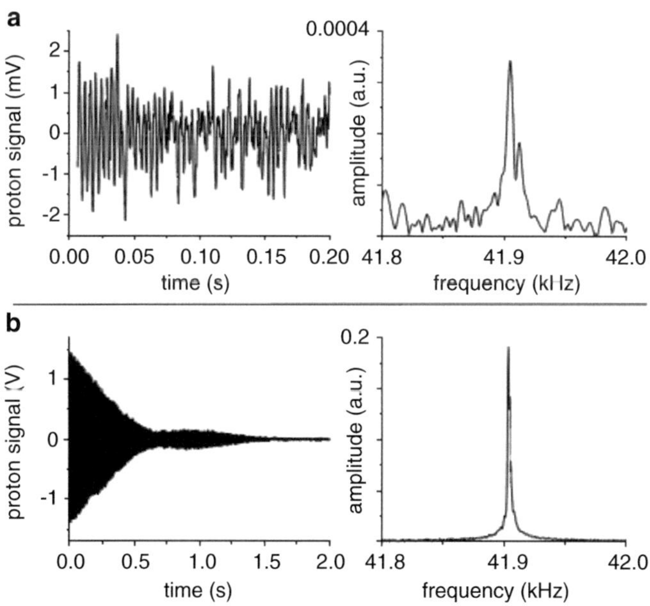

Fig. 4 Comparison of ^{1}H NMR spectra of 0.2 cm^3 toluene. (**a**) Thermal polarization at $B = 9.8 \times 10^{-4}$ T ($\nu = 41.7$ kHz) and 100 scans. (**b**) Hyper-polarization with Xe–^{1}H SPINOE and a single scan at the same magnetic field strength. The ^{1}H NMR signal is enhanced by a factor of 1,000 [31]

2.4 Para-Hydrogen Induced Polarization

PHIP has been used for some decades to enhance the NMR signal during hydrogenation reactions by adding two hydrogen atoms in one molecule at two non-equivalent magnetic positions. Two experiments were known for hydrogenation reactions with a homogeneous catalyst and para-hydrogen. One proceeds at high magnetic field, typically inside a high-field magnet. It is known as Parahydrogen And Synthesis Allow Dramatically Enhanced Nuclear Alignment (PASADENA). The other is performed at low magnetic field outside the magnet and is known as Adiabatic Longitudinal Transport After Dissociation Engenders Net Alignment (ALTADENA). Both differ in the quantum mechanical product operator states of the nuclear spins when the magnetic equivalence of the hydrogen spins is broken upon formation of the chemical transition complex and subsequently leave a different signature in the resultant NMR spectra.

Recently it was discovered that the polarization transfer can proceed even without a chemical reaction as long as a sufficiently long-lived complex between para-hydrogen and target molecules is formed which lifts the magnetic equivalence of the spins in the hydrogen molecule and allows magnetization transfer via the *J*-coupling network of the complex [38, 39]. This technique is called SABRE for Signal Amplification By Reversible Exchange. The outstanding innovation is that the magnetization transfer proceeds without chemical reaction (Fig. 5). This fact offers new possibilities for chemical analysis in particular of small amounts of chemicals which can be polarized selectively at low magnetic fields given a suitable template molecule for formation of the complex.

The first experiments were done on pyridine as a substrate. The substrate molecules exchange with the metal center of a Crabtree catalyst [44], for example $[Ir(COD)(PCy_3)(py)][PF_6]$ (PCy_3: *tris*-cyclohexyl phosphine, py: pyridine, COD: cyclooctadiene), in a solution of methanol. Following the supply of para-hydrogen enriched gas to the solution, an $[Ir(PCy_3)(py_3)(H_2)][PF_6]$ complex forms in which two pyridine molecules are in the same plane as the two hydrogen atoms. The mechanism of polarization transfer at low field is approximated by a conversion of non-observable singlet spin state of para-hydrogen into observable Zeeman magnetization or two-spin alignment in the presence of *J*-coupling and chemical shift

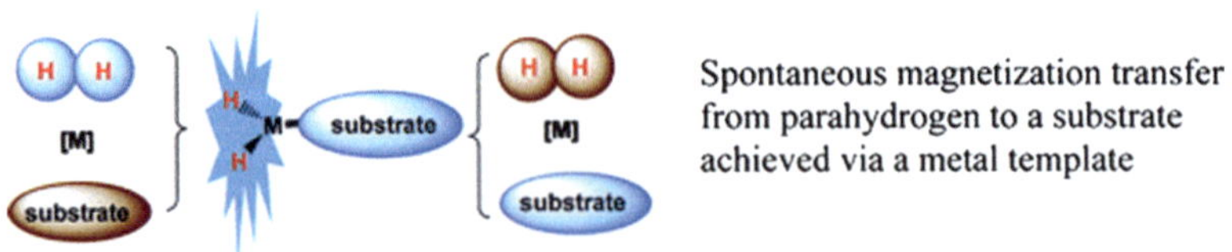

Fig. 5 Illustration of the SABRE-process. A substrate molecule of interest adds to the metal center of a catalyst together with para-hydrogen. Given a sufficiently long lifetime of the complex, polarization is transferred from para-hydrogen to the substrate molecule. The complex subsequently dissociates into its original components. The net effect is a transfer of polarization but no chemical transformation of the substrate molecule [38, 39]

differences. Upon closer examination of the symmetry of the complex it turns out that the coupling of one substrate molecule in *cis*-position to one hydrogen atom is different from the coupling of the same substrate molecule to the other hydrogen which is in *trans*-position.

So far all successful SABRE experiments have used the Crabtree catalyst, and the effect was observed for nitrogen containing substrates only. Furthermore, the experiments were all performed in methanol as the catalyst is insoluble in water. To make use of SABRE for a wider range of molecules including proteins, a water soluble catalyst is needed. Nevertheless, SABRE has opened up new ways for trace analysis, in particular at low magnetic fields. One would expect that the signal-to-noise ratio decreases linearly with the substrate concentration, i.e., the proton concentration. But for SABRE a region was found in which the signal-to-noise ratio increases with decreasing proton concentration of the substrate [41]. At even lower concentrations, a plateau is reached where the signal-to-noise ratio remains constant until the substrate concentration is reduced further by two orders of magnitude. From there on, the expected behavior applies and the signal-to-noise decreases nearly linearly with the substrate concentration. In comparison to hyper-polarization by SABRE, the same substrate shows a linear decrease in the signal-to-noise with decreasing concentration when thermally polarized at $B_p = 2$ T or when hyper-polarized by Xe–^{1}H SPINOE. These effects are illustrated in Fig. 6 for

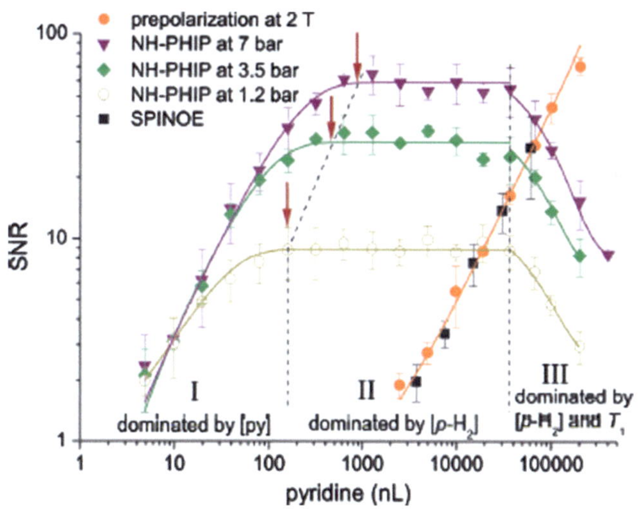

Fig. 6 Dependence of the signal-to-noise ratio in ^{1}H NMR spectroscopy at $B = 3.9 \times 10^{-3}$ T on the pyridine concentration following three polarization methods: 1. thermal polarization, 2. Rb–Xe SPINOE, and 3. SABRE at three different hydrogen pressures. For methods 1 and 2 the signal-to-noise decreases linearly with decreasing pyridine concentration. The *dashed lines* divide the SABRE experiments into three regions. In region *I* the signal-to-noise ratio increases nearly linearly with the pyridine concentration, in region *II* the signal-to-noise ratio is constant and independent of the pyridine concentration, and in region *III* the signal-to-noise ratio decreases with increasing pyridine concentration. The *arrows* indicate where the pyridine concentration is equal to the para-hydrogen concentration [41]

different concentrations of pyridine in methanol-d_4 and three different pressures with para-hydrogen enriched gas.

The pyridine concentration regime can be divided into three regions: region I is dominated by the pyridine concentration, region II by the concentration of para-hydrogen, and region III by the para-hydrogen concentration and the longitudinal relaxation rate T_1 of the protons. The behavior observed in region II can only be explained if the proton T_1-relaxation rate of pyridine and the polarization transfer rate R_p from para-hydrogen to pyridine are constant. In region II the pyridine concentration is so low that intra-molecular relaxation mechanisms dominate, giving a constant proton T_1-relaxation rate of pyridine. If the pyridine concentration is larger than the para-hydrogen concentration, R_p is only proportional to the para-hydrogen concentration (see Fig. 6). In region I the pyridine concentration is lower than the para-hydrogen concentration, and R_p is proportional to the pyridine concentration. As pyridine has a constant proton T_1-relaxation rate, the signal-to-noise ratio drops with the pyridine concentration and therefore one observes a nearly linear decrease if the pyridine concentration is reduced. In region III, where the pyridine concentration is larger than the para-hydrogen concentration, the signal-to-noise ratio decreases with increasing pyridine concentration. This arises from the fact that the proton T_1-relaxation rate of pyridine increases with pyridine concentration due to longer correlation times and the more frequent pyridine–pyridine collisions in the more concentrated pyridine solution. Combining SABRE and the effects described above, it was possible to hyperpolarize a few nanoliters of pyridine in 0.4 cm methanol-d_4 and still observe signal in a single-scan at a magnetic field of $B = 3.9 \times 10^{-3}$ T.

3 Chemical Analysis in the Millitesla Regime and Below

An interesting topic is to investigate the possibilities of chemical analysis at very low field (0–10^{-3} T), in order to delineate the limits imposed by the principles of physics on constructing miniaturized, low-field chemical analyzers. Today, most ^{1}H NMR spectroscopy analysis is performed at high field in the well-known weak coupling regime, where the J-coupling is very small compared to the frequency difference (due to chemical shift differences of one nuclear species or to different nuclear species) between the coupled nuclei, so that the resultant NMR spectra can be explained making use of the familiar rules derived from first-order perturbation theory.

3.1 Measurement of Chemical Shift Differences Below the Line Width

To discuss the requirements for chemical-shift resolved NMR spectroscopy in the millitesla regime [42], where the weak coupling regime is assumed to be valid, two protons A and B with different resonance frequencies ν_A and ν_B are considered. If the frequency difference between ν_A and ν_B is larger than the line width $\Delta\nu = 1/(\pi T_2^*)$ the two lines corresponding to chemical groups A and B can be resolved. The free induction signal decays according to $1/T_2^* = 1/T_2 + \gamma\Delta B_0(\boldsymbol{r})$. The effective relaxation rate depends on the transverse relaxation time T_2, the gyromagnetic ratio γ and the field inhomogeneity $\Delta B_0(\boldsymbol{r})$ across the sample volume. As the two different resonance frequencies depend on the magnetic field B and on the chemical shifts δ_A and δ_B in such a way, that $\nu_A = \gamma\, B\, (1 + \delta_A)$ and $\nu_B = \gamma\, B\, (1 + \delta_B)$, one can conclude in view of the assumption $|\nu_A - \nu_B| > \Delta\nu$ that the magnetic field needed to resolve the two lines must obey $B > \Delta\nu/[\gamma\, (\delta_A - \delta_B)]$. For example for ^{1}H NMR and in the absence of hetero nuclear J-coupling, a chemical shift difference of 10 ppm between two ^{1}H lines A and B with line width $\Delta\nu = 0.4$ Hz could be resolved at $B = 10^{-3}$ T. This indeed corresponds to the example shown in Fig. 2c with the ^{1}H NMR spectrum of acetic acid acquired at $B = 9.8 \times 10^{-4}$ T (41.7 kHz). An open-gap electromagnet magnet without shims, consisting of two large solenoid coils provided enough homogeneity (~3 ppm/cm) to detect the chemical shift difference of 9.8 ppm (Fig. 2) between the ^{1}H-signals of the methyl and the acid groups. In this case, the line width limits the chemical shift resolution. But in certain cases, this line width limitation can be circumvented, e.g., if a magnetic hetero-nucleus like ^{13}C or ^{29}Si couples to the protons. The hetero-nuclear J-coupling splits at least one of the peaks into a multiplet, preferably a doublet. Consider two chemical groups A and B with two lines at frequency ν_A and ν_B. We assume that the frequency difference $|\nu_A - \nu_B|$ is smaller than the line width $\Delta\nu$ (Fig. 7a, left). The chemical shift difference can be determined if group A couples to a hetero-nucleus with spin 1/2. This coupling leads to a splitting of the line at frequency ν_A into two lines at frequencies ν_{A1} and ν_{A2} where $|\nu_{A2} - \nu_{A1}| = J$ (Fig. 7a, right).

The positions of the maxima of these lines can be used to determine the frequency difference $\nu_A - \nu_B = (\nu_{A2} - \nu_{A1})/2 - \nu_B$ with a precision that is much higher than the line width $\Delta\nu$. For example, the ^{1}H chemical shift difference between the methyl and the hydroxyl groups in methanol (Fig. 7b, left) is 1.5 ppm or 0.24 Hz at $B = 3.9 \times 10^{-3}$ T. This is less than the line width of $\Delta\nu - 0.4$ Hz, so that the chemical shift cannot be determined in a straightforward way. But in ^{13}C enriched methanol the hetero-nuclear coupling $^1J_{H,C}$ between the methyl protons and the ^{13}C leads to a splitting of the methyl line into two lines separated by $^1J_{H,C} = 140$ Hz. The chemical shift difference between the hydroxyl and the methyl groups can be determined from the center between the maxima of the methyl doublet and the maximum of the hydroxyl group line. The observed difference in chemical shift is indeed 1.5 ppm (0.24 Hz). By combining this approach with signal enhancement by pre-polarization or hyper-polarization and with more sensitive detection methods (using e.g., atomic magnetometers) one may also determine chemical shift differences below the resolution limit imposed by the line width by making use of the satellite peaks of ^{1}H NMR spectra with ^{13}C in natural abundance.

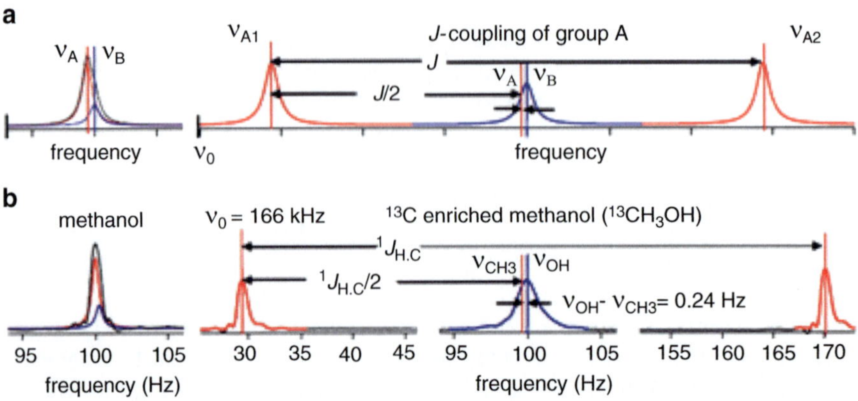

Fig. 7 (**a**) Simulation of the spectrum of two chemical sites at frequencies ν_A and ν_B, without (*left*) and with (*right*) hetero-nuclear *J*-coupling of the species A. (**b**) ^{1}H spectrum of methanol without (*left*) and with (*right*) ^{13}C enrichment at $B = 3.9 \times 10^{-3}$ T (166 kHz ^{1}H frequency). In the presence of the hetero-nuclear *J*-coupling it is possible to determine the chemical shift difference of 1.5 ppm between the hydroxyl and the methyl groups [42]

3.2 *Weak and Strong J-Coupled NMR-Spectroscopy in Magnetic Fields from 0 to 100 T*

As the magnetic field strength is decreased, the strong coupling regime is reached where *J*-couplings and frequency differences between the nuclei become comparable. Upon entering this regime, the NMR spectra become more and more complex up to a turning point from where on the spectra become simpler again. As it turns out, spectra at ultra-low magnetic fields follow a dual correspondence with spectra at high-field in terms of their simplicity. The following section discusses how chemical analysis can be performed at ultra-low magnetic fields down to 0 T.

In the following, the question is addressed as to whether NMR spectra provide sufficient information for chemical analysis when going down to zero-field. In particular, the structure of *J*-coupled NMR spectra close to zero-field is investigated, and an answer is sought to the question if one can retrieve structural information in the absence of chemical shift information. To this end, a model is discussed that postulates the structure of a *J*-coupled spectrum from very high magnetic fields down to zero field [45]. The spectra for an SI_3 spin system in the different regimes from ultra-high to ultra-low magnetic fields are depicted in Fig. 8.

In very high magnetic fields and in the standard high field NMR regime, the Hamiltonian is dominated by the Zeeman interaction and the *J*-coupling term acts as a perturbation. Close to zero-field an inverse behavior is observed that leads to a different but simple pattern in the NMR spectrum. Here, the *J*-coupling is the dominating term in the Hamiltonian and the Zeeman interaction can be understood as a perturbation. Between these regions lie different regimes that can be approximated by different orders of perturbation theory and that show very

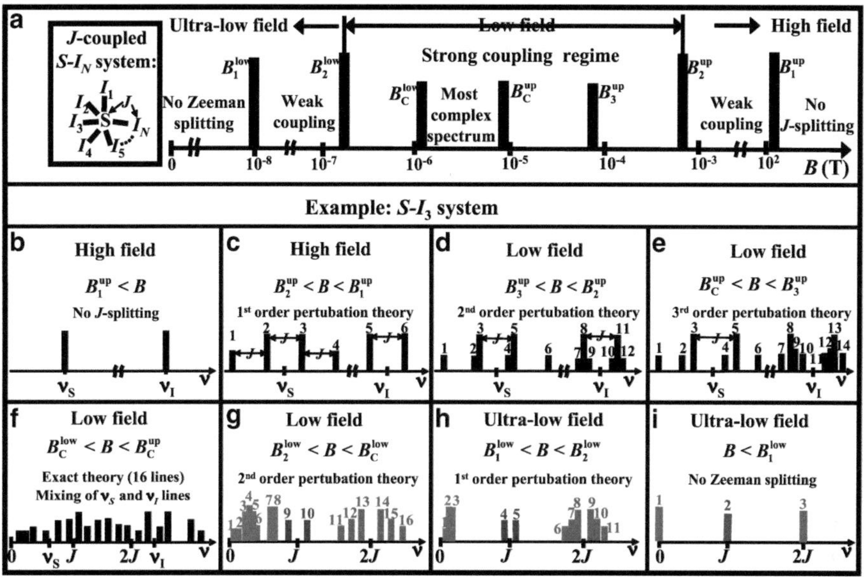

Fig. 8 (**a**) Strong and weak coupling regimes of the S–I_3 system as a function of magnetic field B. (**b**) Stick spectrum at very high field $B > B_1^{\mathrm{up}}$. (**c**) Weak coupling regime $B_2^{\mathrm{up}} < B < B_1^{\mathrm{up}}$. (**d–g**) Strong coupling regimes $B_2^{\mathrm{low}} < B < B_2^{\mathrm{up}}$. (**h**) Inverse weak coupling regime $B_1^{\mathrm{low}} < B < B_2^{\mathrm{low}}$. (**i**) Close to zero field $B < B_1^{\mathrm{low}}$ [45]

complicated NMR spectra. We start considering an S–I_3 group, where $\nu_{\mathrm{I}} = \gamma_{\mathrm{I}} B$, $\nu_{\mathrm{S}} = \gamma_{\mathrm{S}} B$ are the Larmor frequencies of spins I and S with their corresponding gyromagnetic ratios γ_{I}, and γ_{S}, and a hetero-nuclear coupling constant J in a magnetic field B, that is not perfectly homogeneous. Both coupled spins possess a nuclear spin of 1/2 and the magnetic field inhomogeneity is 25 ppb. At ultra-high magnetic fields larger than $B_1^{\mathrm{up}} = 100\mathrm{T}$ (corresponding to a proton Larmor frequency of >4.26 GHz) the corresponding ^{1}H NMR line width is $\Delta\nu > 100$ Hz. Therefore assuming $J = 100$ Hz the spectrum – or rather the two spectra for each type of nucleus – would show only two lines at frequency ν_{I} and ν_{S} and no splitting due to the J-coupling. So for $B > B_1^{\mathrm{up}}$ we have $\Delta\nu > J$, and the only information in the NMR spectrum is that two nuclear spin species I and S are present (Fig. 8b), and their chemical shift can be determined with unsurpassed precision.

In the range of $B_2^{\mathrm{up}} < B < B_1^{\mathrm{up}}$, where $B_2^{\mathrm{up}} = J^2/[2\Delta\nu(\gamma_{\mathrm{I}} - \gamma_{\mathrm{S}})]$, (i.e., the J-coupling is larger than the line width but smaller than the difference in Larmor frequencies of both nuclei) the weak coupling regime in high magnetic fields is entered, a region in which common NMR spectrometers are operating. In this regime the spectra can be explained by first order perturbation theory. The J-coupled spectrum of the S–I_3 group in this regime is an I-spin doublet an S-spin quartet (Fig. 8c). The spectrum contains information about the spin species, the J-coupling constant, and the number of spins involved. Typically for a $^{13}CH_3$ group, with $J = 140$ Hz and a line width $\Delta\nu = 0.33$ Hz ($T_2 = 1\mathrm{s}$),

$B_2^{up} = 10^{-3}$T, meaning that a weakly coupled standard NMR spectrum is measured if B is larger than 10^{-3} T (corresponding to a proton Larmor frequency of 42.6 kHz).

At still lower external magnetic fields, in the range $B_3^{up} < B < B_2^{up}$ the upper strong coupling limit is entered, and the value for B_3^{up} is given by $B_3^{up} = (3J^3/[4\Delta\nu(\gamma_I - \gamma_S)^2])^{1/2}$. For a $^{13}CH_3$ group and assuming a line width of $\Delta\nu = 0.33$ Hz and $J = 140$ Hz we obtain $B_3^{up} = 8.3 \times 10^{-5}$T, corresponding to a proton Larmor frequency of 3.5 kHz. In this upper strong coupling regime the observed ^{1}H spectrum can be explained by a degenerate vector model. If one proton of the $^{13}CH_3$ group is observed, the two other protons can couple into a triplet state, with magnetic quantum numbers 1, 0, and −1 or into a singlet state. As the energy of the zero-state of the triplet and of the singlet state are degenerate in the discussed limit, three separate lines occur (Fig. 8d). Therefore, the ^{1}H spectrum consists of a pair of three lines, and the central lines of both triplets are separated by J. On the other hand, if the ^{13}C nucleus is observed, the three protons can strongly couple to a total spin of 3/2 (with four magnetic states) and to 1/2 (with two magnetic states) giving rise to six separate lines. In conclusion, for $B_3^{up} < B < B_2^{up}$ the total spectrum of an S–I_3 group consists of six peaks around ν_S and a pair of triplets around ν_I (see Fig. 8d).

The next strong coupling regime is found to be $B_C^{up} < B < B_3^{up}$ (see Fig. 8e) where $B_c^{up} = (1 + (1/2)^{1/2})J/(\gamma_I - \gamma_S)$. The earth's magnetic field is inside this regime, and the earth's field spectrum is more complicated compared to the spectrum at $B_3^{up} < B < B_2^{up}$, and a non-degenerate vector model is needed to explain its features. In the following this vector model is presented for the S–I_N group and especially for the case $^{13}CH_3$ [47]. In high magnetic fields only the projection of nuclear spins in the direction of the field B, the z-component, is relevant. As the magnetic field decreases, x- and y-components of nuclear spins can no longer be neglected, and components in all three directions have to be taken into account. Let us first consider the $^{13}CH_3$ group. For the protons two sets of four lines separated by 140 Hz are observed, arising from a triplet and a singlet state, where the zero state of the triplet and the singlet are no longer degenerate. These lines can be explained in the following way: If one of the protons is observed, the two others can couple strongly to a total spin $L = 1$, giving a triplet, and to a total spin $L = 0$, resulting in a singlet. Compared to the regime $B_3^{up} < B < B_2^{up}$ the singlet and zero triplet state are not degenerate any longer. Considering the carbon spectrum of the $^{13}CH_3$ group in the field regime $B_C^{up} < B < B_3^{up}$, six lines can be derived because, as described above, the three protons couple together to a total spin of 3/2 or 1/2. Therefore the whole spectrum of an S–I_3 group consists of 14 peaks, a pair of four lines grouped around ν_I and six lines grouped around ν_S (Fig. 8e).

Figure 9 summarizes the ^{1}H spectra for a $^{13}CH_3$ group of methanol in the weak coupling regime, and two upper strong coupling regimes. Measured spectra are compared to simulated spectra, showing a perfect match and proving the existence of the predicted boundaries B_3^{up} and B_2^{up}. In the weak coupling regime only two lines for the protons, separated by $^1J_{H,C}$, are observed (Fig. 9a). At $B = 9.88 \times 10^{-4}$ T

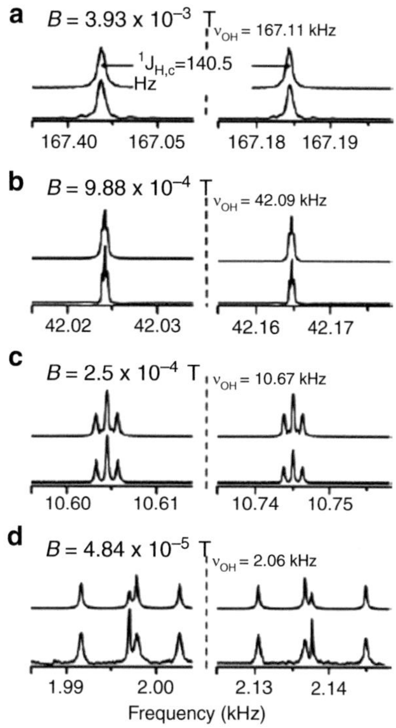

Fig. 9 Simulated and measured evolution of the ^{13}C enriched ^{1}H methanol spectrum if the boundary between weak and strong coupling limits is transgressed. (**a**) $B = 3.9 \times 10^{-3}$ T still lies in the weak coupling regime. (**b**) At $B = 9.88 \times 10^{-4}$ T the doublet starts to split into a pair of three lines. In (**c**) $B = 2.5 \times 10^{-4}$ T and compared to (**b**) the splitting between the *three lines* is larger. In (**d**) $B = 4.8 \times 10^{-5}$ T $< B_3^{up} = 8.3 \times 10^{-5}$ T and the spectrum splits further into a pair of *four lines* [45]

the strong coupling regime is entered and the degeneracy of the singlet and triplet zero state become observable in a pair of three lines (Fig. 9b). If the static magnetic field is reduced further the pair of three lines become more obvious at $B = 2.5 \times 10^{-4}$ T (Fig. 9c). Finally, in the earth's magnetic field the degeneracy between the singlet and the triplet zero state is lifted giving rise to a pair of four lines (Fig. 9d). A remarkable observation is that the line which is associated with the singlet state of the non-observed two protons is much narrower than the other three lines arising from the triplet state [46, 47].

Generalizing to an S–I_N group every line of the I-doublet spaced by J in the upper weak coupling regime will split in the strong coupling regime $B_C^{up} < B < B_3^{up}$ into $K = \sum_{n=1,n\varepsilon U}^{N} (N - n + 1)$ lines, where U is the set of odd numbers and n is a natural number. For example, the ^{1}H spectrum of a strongly coupled SH_N group reveals for $N = 1, 2, 3, 4, 5$ a pair of one, two, four, six, and nine lines, a behavior which can be explained by the vector model as depicted in Fig. 10a–e.

For the S–I group an I-spin doublet is observed, with both peaks separated by J (Fig. 10a) If there are N spins I and one S spin, the $(N-1)$ unobserved I-spins couple to a total spin L with $L = (N-1)/2, (N-3)/2, (N - 5)/2,\ldots, (N - n)/2$ with n ε U down to $N - n = 0$ for odd N or to $N - n = 1$ for even N. Every total spin L has $2L + 1$ possible orientations giving the K peaks described above if summed up over all orientations (Fig. 10b–e). Figure 10f, g shows what happens in the earth field, if two molecular groups S–I_N and A–I_{M-N} exist and (M–N) I-spins are

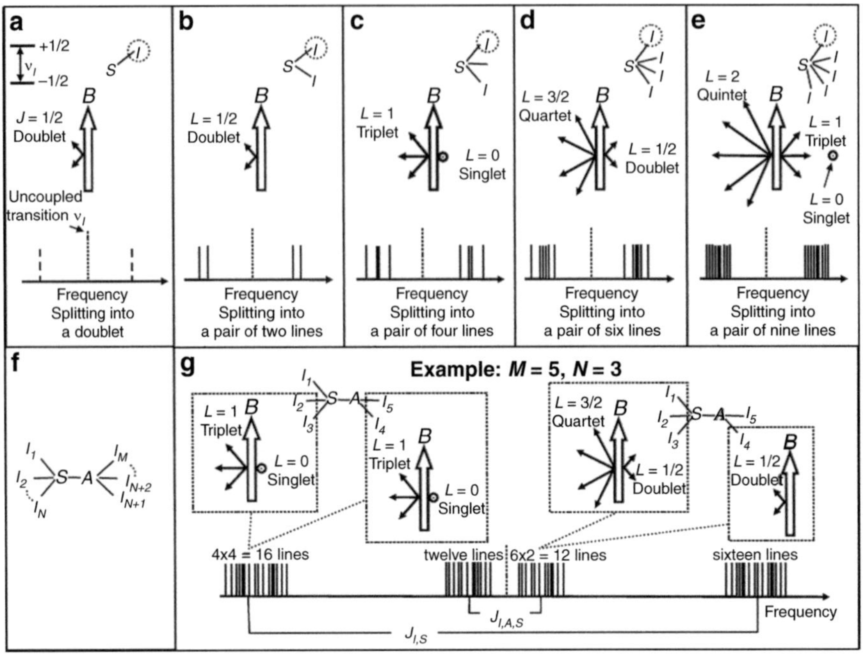

Fig. 10 Non-degenerate vector model to determine the multiplicity in the upper strong coupling regime $B_C^{up} < B < B_3^{up}$ for a hetero-nuclear *J*-coupled S–I_N group (**a**–**e**) and for a homo- and hetero-nuclear *J*-coupled molecule I_3S–AI_2 (**f, g**) [47]

homo-nuclear *J*-coupled to N *I*-spins of the other group. As a special case, Fig. 10g depicts the spectrum for the spin system I_3S–AI_2. If nucleus I_1 is observed, nuclei I_2 and I_3 as well as I_4 and I_5 couple to a total spin. Thereby $L_{I2I3} = (3 - 1)/2 = 1$ and $L_{I2I3} = (3 - 3)/2 = 0$, giving a triplet and a singlet. The total spin of I_4 and I_5 is $L_{I4I5} = (3 - 1)/2 = 1$ and $L_{I4I5} = (3 - 3)/2 = 0$, which again results in a triplet and a singlet. Therefore, four lines are further split into four lines each, resulting in 16 lines in total. On the other hand, if I_5 is observed a total spin of I_1, I_2, and I_3 is established and a total spin of I_4, giving $L_{I1I2I3} = (4 - 1)/2 = (3/2)$ and $L_{I1I2I3} = (4 - 3)/2 = (1/2)$, which is a quartet and a doublet. $L_{I4} = (2 - 1)/2$ also results in a doublet. All peaks couple again with each other and 12 lines are observed. Thus the whole spectrum of the described molecule consists of 2×16 and 2×12 lines, which corresponds to an ethanol molecule if one carbon is ^{13}C enriched. The OH-group in ethanol is not involved in the *J*-coupled network due to its mobility and gives rise to one uncoupled line at the Larmor frequency.

Figure 11 shows a measured ^1H earth's field spectrum of ethanol, ^{13}C enriched at the methyl group (HO–CH_2–$^{13}CH_3$). The observed lines correspond to the I_3S–AI_2 molecule as is confirmed by a simulation (Fig. 11) based on the vector model. In these spectra the homo-nuclear *J*-couplings between the protons can be observed because different hetero-nuclear *J*-couplings between ^{13}C–H and H–^{12}C–^{13}C break the magnetic equivalence of the protons in the CH_2 and CH_3 groups [46].

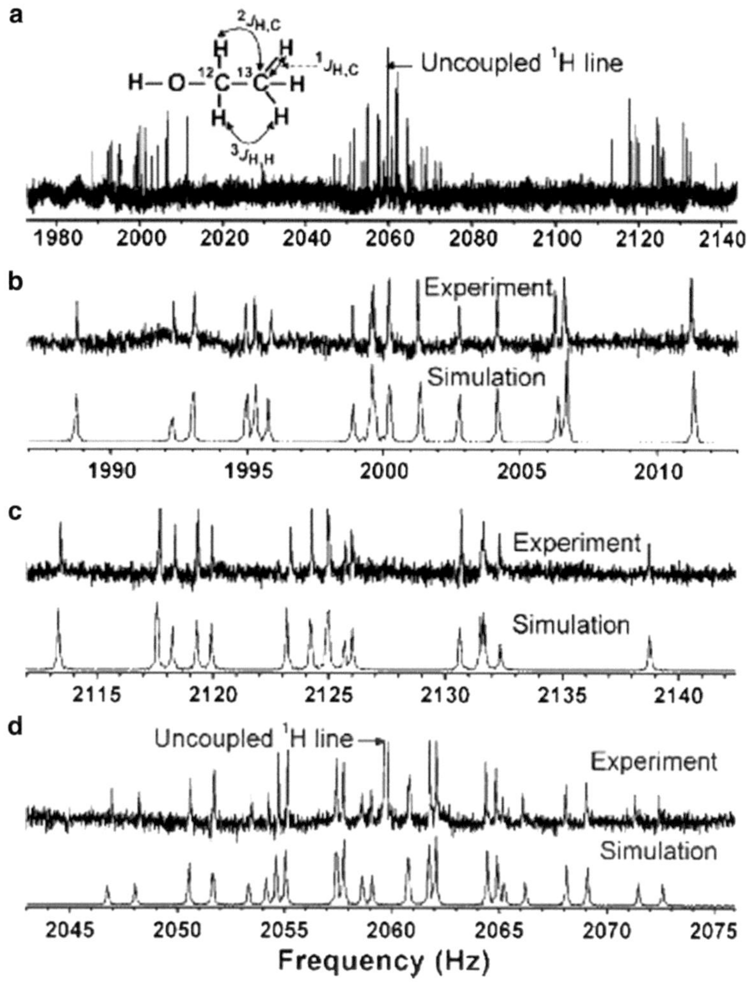

Fig. 11 *J*-coupled earth's field ^{1}H spectrum of ethanol, which is ^{13}C enriched at the CH_3 group. (**a**) The complete spectrum with 2 × 16 and 2 × 12 lines. (**b**, **c**) Enhanced view of the outer 16 lines at each side of the spectrum. (**d**) The inner 2 × 12 lines and the uncoupled OH proton line. Underneath the experimental data, spectra simulated using the vector model are shown, which confirm the notion of the multiplicity [47]

The dynamics of OH exchange is shown in Fig. 12 by example of the earth's field spectrum of $^{13}CH_3$–OH [46]. As long as the protons of the OH group are mobile, for example at 60 °C (Fig. 12a), there is no homo-nuclear *J*-coupling observable, and the investigated methanol molecule corresponds to a strongly coupled S–I_3 system in the earth's magnetic field so that 2 × 4 peaks are observed plus a rapidly exchanging OH peak without coupling. At a temperature of −80 °C, which is still above the melting point of methanol, the OH proton is bound permanently to the methanol molecule. Due to the different hetero-nuclear

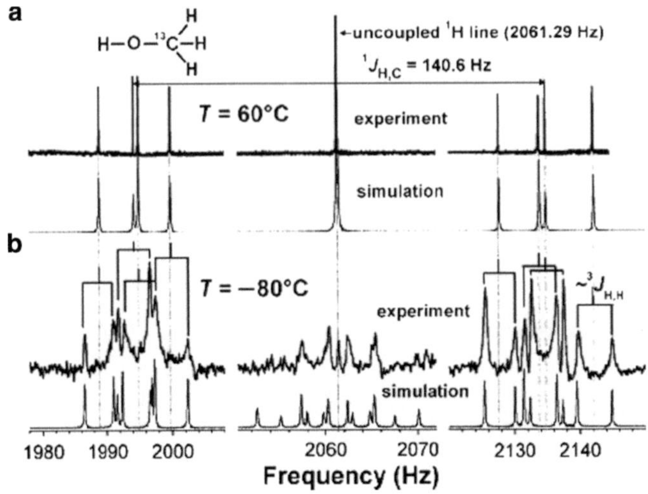

Fig. 12 ^{1}H NMR spectra of a ^{13}C enriched methanol sample acquired with a single scan in the earth's magnetic field at different temperatures to demonstrate the influence of molecular dynamics. At 60°C (**a**) the $^{13}CH_3$ group corresponds to the case shown in Fig. 10c, which results in a pair of *four lines*. At −80°C (**b**) the OH proton is attached to the molecule and couples to the methyl group. The homo-nuclear *J*-coupling between the protons of the OH and the $^{13}CH_3$ groups can be observed [46]

J-couplings between ^{13}C–OH and $^{13}C–H_3$, the magnetic equivalence between the OH and the CH_3 protons is broken, and the homo-nuclear coupling $^{3}J_{H,H}$ can be observed (Fig. 12b). Formally, the $S–I_3$ group is transformed into an $I_3–S–AI$ system by cooling, and the total number of observed peaks is 2 × 8 plus 2 × 6.

If the magnetic field is lowered further, a region with most complex spectra is reached between $B_c^{low} \approx 4J/(7\gamma_I + \gamma_S)$ and B_c^{up}, in which the maximal number of $(N + 1)$ lines can be observed for a strongly coupled $S–I_N$ group. For a $^{13}CH_3$ group, that means 16 peaks (Fig. 8f) are observable. The non-degenerate vector model predicts the existence of 14 lines, 6 lines around the carbon frequency and 8 lines around the proton frequency. The two additional peaks are combination lines of three-spin transitions, in which two protons and one carbon simultaneously change their orientation. This three spin transition occurs while the third proton of the $^{13}CH_3$ group is either in the up or down position.

In the range of $B_2^{low} = (2\Delta\nu J/(\gamma_I - \gamma_S)^2)^{1/2} < B < B_c^{low}$, where $B_2^{low} = 3\times 10^{-7}$T (a magnetic field strength that would correspond to a proton Larmor frequency of 12.8 Hz), 16 lines are still observable for a $^{13}CH_3$ group but they start grouping around the frequencies 0, *J* and 2*J* (Fig. 8g). Between the limits $B_1^{low} = \Delta\nu/(\gamma_I - \gamma_S) < B < B_2^{low}$ a regime is reached, in which the spectra have a simplicity comparable to that encountered in the high-field regime with weak coupling. The observed number of lines for the $S–I_3$ group is 11 (Fig. 8h), 2 × 3 lines around frequency 2*J*, 2 lines around frequency *J* and 3 lines around 0. For the $S–I_N$ spin system simple multiplet structures group around the frequencies 0, *J*, (3/2)*J*, 2*J*,

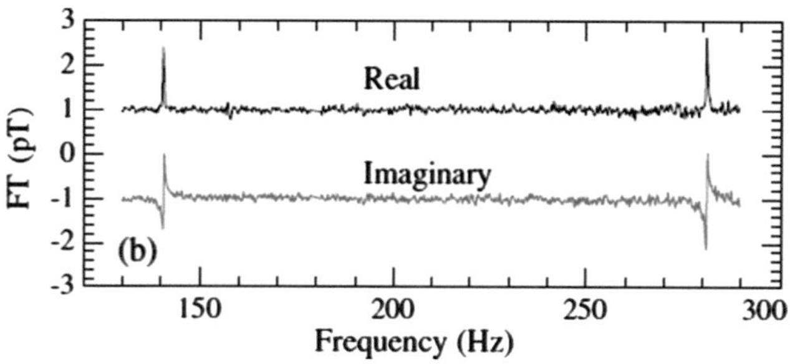

Fig. 13 Fourier-transformed spectrum of ^{13}C enriched methanol $^{13}CH_3OH$. $^1J_{H,C}$ in methanol equals 140 Hz and *two lines* are observed at J and $2J$ [8]

(5/2)J, etc. The number and frequency positions of these multiplet lines correspond unambiguously to the number of I nuclei coupled to the S nucleus and can be explained by inverse first-order perturbation theory. Inverse first-order perturbation means that the Zeeman energy is much smaller than the energy splitting due to the J-coupling, so the Zeeman Hamiltonian acts as the perturbation. This region is called the inverse weak coupling regime and is in dual correspondence to the high-field weak coupling regime. For both the upper and lower weak coupling regimes the J-coupled spectrum allows an unambiguous determination of the spin species, the J-coupling constant, and the number of coupled spins.

Below B_1^{low}, which is around 10^{-8} T, down to zero-field a region (Fig. 8i) with simple spectra in dual correspondence to the regime at ultra-high field (Fig. 7b) is reached. Here, single peaks are encountered centered at frequencies 0, J, (3/2)J, 2J, (5/2)J, etc. Figure 8b, i are in dual correspondence in the following sense: Fig. 8i has J-coupling information but no information about the spin species and the number of spins involved while Fig. 8b bears the information on the spin species but no information about the number of spins and the J-coupling.

Recently it has been shown that J-coupled spectra can be acquired with atomic magnetometers at zero magnetic field, also resolving homo-nuclear J-couplings, which leads to a splitting around the positions 0, J, (3/2)J, 2J, (5/2)J, etc. In Fig. 13, a measured spectrum at zero field is shown for a $^{13}CH_3$ group, with two lines at J and at 2J [8].

The problem encountered when attempting to perform chemical analysis at zero magnetic field with the S–I_N group is that the spectra are not unambiguous. Figure 14 shows spectra of different chemical groups at zero field and at 2×10^{-7} T.

At zero field it is not possible to distinguish between a ^{13}CH and a $^{15}NH_2$ group, whereas at 2×10^{-7} T different multiplet structures are observed. From simulations and measurements in the strong coupling regime as well as at zero-field it is postulated that in future chemical analysis can be done in the inverse weak coupling regime $B_1^{\mathrm{low}} < B < B_2^{\mathrm{low}}$ because sufficient information to determine unambiguously structures of chemical groups is present, although the complexity for many coupled chemical groups is high. The small field of about 10^{-7} T

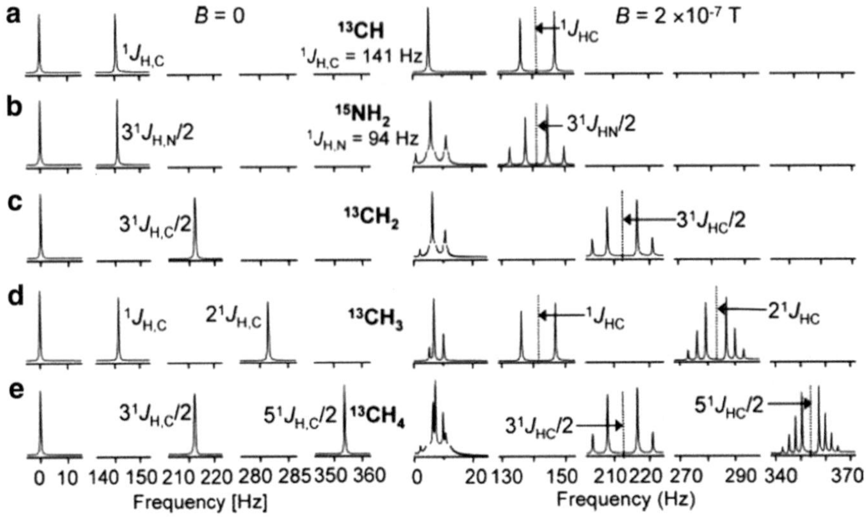

Fig. 14 Simulated *J*-coupled NMR spectra near zero field for different molecular groups (**a**) ^{13}CH, (**b**) $^{15}NH_2$, (**c**) $^{13}CH_2$, (**d**) $^{13}CH_3$, and (**e**) $^{13}CH_4$. Spectra are shown at zero-field (*left*) and in the lower weak coupling regime at 2×10^{-7} T (*right*). The number of lines and their separations measured at $B = 2 \times 10^{-7}$ T determine the SI_N groups in a unique way [45]

necessary for assigning the chemical groups does not need to be homogeneous. Homogeneity of one part per thousand would be sufficient for the measurement of highly resolved spectra. The determination of the structure of large molecules, such as proteins, is still unclear, but NMR at ultra-low field certainly opens up new opportunities for chemical analysis of molecules without superconducting magnets.

References

1. Perlo J, Demas V, Casanova F, Meriles C, Reimer J, Pines A, Blümich B (2005) High-resolution NMR spectroscopy with a portable single-sided sensor. Science 308:1279
2. Perlo J, Casanova F, Blümich B (2007) Ex situ NMR in highly homogeneous fields: ^1H spectroscopy. Science 315:1110–1112
3. McDermott R, Trabesinger AH, Mück M, Hahn EL, Pines A, Clarke J (2002) Liquid-state NMR and scalar couplings in microtesla magnetic fields. Science 295:2247–2249
4. Appelt S, Häsing FW, Kühn H, Perlo J, Blümich B (2005) Mobile high resolution xenon nuclear magnetic resonance spectroscopy in the earth's magnetic field. Phys Rev Lett 94:197602
5. Appelt S, Kühn H, Häsing FW, Blümich B (2006) Chemical analysis by ultrahigh-resolution nuclear magnetic resonance in the Earth's magnetic field. Nat Phys 2:105–109
6. Robinson JN, Coy A, Dykstra R, Eccles CD, Hunter MW, Callaghan PT (2006) Two-dimensional NMR spectroscopy in Earth's magnetic field. J Magn Reson 182:343–347
7. Savukov IM, Romalis MV (2005) NMR detection with an atomic magnetometer. Phys Rev Lett 94:123001

8. Ledbetter MP, Crawford CW, Pines A, Wemmer DE, Knappe S, Kitching J, Budker D (2009) Optical detection of NMR J-spectra at zero magnetic field. J Magn Reson 199:25–29
9. Packard M, Varian R (1954) Free nuclear induction in the Earth's magnetic field. Phys Rev 93:941
10. Benoit H, Hennequin J, Ottavi H (1962) Les applications spectroscopiques de la méthode de prépolarisation en R.M.N. (champs faibles). Chim Anal 44:471–477
11. Béné GJ (1980) Nuclear magnetism of liquid systems in the Earth field range. Phys Rep 58:213–267
12. Matlachov AN, Volegov PL, Espy MA, George JS, Kraus RH Jr (2004) SQUID detected NMR in microtesla magnetic fields. J Magn Reson 170:1–7
13. Burghoff M, Hartwig S, Trahms L, Bernarding J (2005) Nuclear magnetic resonance in the nano tesla range. Appl Phys Lett 87:054103
14. Savukov IM, Lee S-K, Romalis M (2006) Optical detection of liquid-state NMR. Nature 442:1021–1024
15. Budker D, Romalis M (2007) Optical magnetometry. Nature Phys 3:227–234
16. Balasubramanian G, Chan IY, Kolesov R, Al-Hmoud M, Tisler J, Shin C, Kim C, Wojcik A, Hemmer PR, Krueger A, Hanke T, Leitenstorfer A, Bratschitsch R, Jelezko F, Wrachtrup J (2008) Nanoscale imaging magnetometry with diamond spins under ambient conditions. Nature 455:648–651
17. Maze JR, Stanwix PL, Hodges JS, Hong S, Zaylor JM, Cappellaro P, Jiang L, Gurudev Dutt MV, Togan E, Zibrov AS, Yacoby A, Walsworth RL, Lukin MD (2008) Nanoscale magnetic sensing with an individual electronic spin in diamond. Nature 455:644–647
18. Kleinberg RL (1996) Well logging. In: Encyclopedia of NMR. Wiley-Liss, New York, pp 4960–4969
19. Coates GR, Xiao L, Prammer MG (1999) NMR logging – principles and applications. Halliburton Energy Services, Houston
20. Kimmich R, Anoardo E (2004) Field-cycling NMR relaxometry. Prog Nucl Magn Reson Spectrosc 44:257–320
21. Noack F (1986) NMR field-cycling spectroscopy: principles and applications. Prog Nucl Magn Reson Spectrosc 18:171–276
22. Callaghan PT, Coy A, Dykstra R, Eccles CD, Halse ME, Hunter MW, Mercier OR, Robinson JN (2007) New Zealand developments in Earth's field NMR. Appl Magn Reson 32:63–74
23. Thayer AM, Pines A (1987) Zero-field NMR. Acc Chem Res 20:47–53
24. Ivanov D, Redfield AG (2004) Field-cycling method with central transition readout for pure quadrupole resonance detection in dilute systems. J Magn Reson 166:19–27
25. Abragam A, Goldman M (1978) Principles of dynamic nuclear polarisation. Rep Prog Phys 41:395–467
26. Halse ME, Callaghan PT (2008) A dynamic nuclear polarization strategy for multi-dimensional Earth's field NMR. J Magn Reson 195:162–168
27. Lingwood MD, Ivanov IA, Cote AR, Han S (2010) Heisenberg spin exchange effect of nitroxide radicals on Overhauser dynamic nuclear polarization in the low field limit at 1.5 mT. J Magn Reson 204:56–63
28. Happer W (1972) Optical pumping. Rev Mod Phys 44:169–249
29. Appelt S, Baranga AB, Erickson CJ, Romalis MV, Young AR, Happer W (1998) Theory of spin-exchange optical pumping of ^{3}He and ^{128}Xe. Phys Rev A 58:1412–1439
30. Navon G, Song YQ, Room T, Appelt S, Taylor RE, Pines A (1996) Enhancement of solution NMR and MRI with laser-polarized xenon. Science 271:1848–1851
31. Appelt S, Häsing FW, Baer-Lang S, Shah NJ, Blümich B (2001) Proton magnetization enhancement of solvents with hyperpolarized xenon in very low magnetic fields. Chem Phys Lett 348:263–269
32. Bowers CR, Weitekamp DP (1986) Transformation of symmetrization order to nuclear-spin magnetization by chemical reaction and nuclear magnetic resonance. Phys Rev Lett 57:2645–2648

33. Natterer J, Bargon J (1997) Parahydrogen induced polarization. Prog Nucl Magn Reson Spectrosc 31:293–315
34. Hubler P, Giernoth R, Kummerle G, Bargon J (1999) Investigating the kinetics of homogeneous hydrogenation reactions using PHIP NMR spectroscopy. J Am Chem Soc 121:5311–5318
35. Aime S, Canet D, Dastru W, Gobetto R, Reineri F, Viale A (2001) A novel application of p-H_2: the reversible addition/elimination of H_2 at a Ru_3 cluster revealed by the enhanced NMR emission resonance from molecular hydrogen. J Phys Chem A 105:6305–6310
36. Bouchard LS, Burt SR, Anwar MS, Kovtunov KV, Koptyug IV, Pines A (2008) NMR imaging of catalytic hydrogenation in microreactors with the use of para-hydrogen. Science 319:442–445
37. Kovtunov KV, Beck IE, Bukhtiyarov VI, Koptyug IV (2008) Observation of parahydrogen-induced polarization in heterogeneous hydrogenation on supported metal catalysts. Angew Chem 120:1514–1517
38. Adams RW, Aguilar JA, Atkinson KD, Cowley MJ, Elliott PIP, Duckett SB, Green GGR, Khazal IG, Lopez-Serrano J, Williamson DC (2009) Reversible interactions with para-hydrogen enhance NMR sensitivity by polarization transfer. Science 323:1708–1711
39. Atkinson KD, Cowley MJ, Elliott PIP, Duckett SB, Green GGR, Lopez-Serrano J, Whitwood AC (2009) Spontaneous transfer of parahydrogen derived spin order to pyridine at low magnetic field. J Am Chem Soc 131:13362–13368
40. Bowers CR, Weitekamp DP (1987) Parahydrogen and synthesis allow dramatically enhanced nuclear alignment. J Am Chem Soc 109:5541–5542
41. Gong Q, Gordji-Nejad A, Blümich B, Appelt S (2010) Trace analysis by low-field NMR: breaking the sensitivity limit. Anal Chem 82:7078–7082
42. Appelt S, Glöggler S, Häsing FW, Sieling U, Gordji-Nejad A, Blümich B (2010) NMR spectroscopy in the milli-tesla regime: measurement of ^1H chemical-shift differences below the line width. Chem Phys Lett 485:217–220
43. Glöggler S, Blümich B, Appelt S (2011) Real-time detection of polymerization reactions with hyperpolarized xenon at low magnetic fields. AIP Conf Proc 1330:101–104
44. Crabtree RH, Lavin M, Bonneviot L (1986) Some molecular hydrogen complexes of iridium. J Am Chem Soc 108:4032–4037
45. Appelt S, Häsing FW, Sieling U, Gordji-Nejad A, Glöggler S, Blümich B (2010) Paths from weak to strong coupling in NMR. Phys Rev A 81:023420
46. Appelt S, Häsing FW, Kühn H, Sieling U, Blümich B (2007) Analysis of molecular structures by homo- and hetero-nuclear J-coupled NMR in the ultra-low field. Chem Phys Lett 440:308–313
47. Appelt S, Häsing FW, Kühn H, Sieling U, Blümich B (2007) Phenomena in *J*-coupled nuclear magnetic resonance spectroscopy in low magnetic fields. Phys Rev A 76:023420

Top Curr Chem (2013) 335: 23–70
DOI: 10.1007/128_2011_229

Published online: 25 October 2011

Dynamic Nuclear Hyperpolarization in Liquids

Ulrich L. Günther

Abstract Nuclear magnetic resonance (NMR) spectroscopy is a broadly used analytical method with major applications in chemistry, biochemistry and medicine. Key applications include structural analysis of small molecules, metabolites, larger biomolecules such as proteins, RNA and DNA, and applications in material science. Magnetic resonance imaging (MRI), which is based on the same physical principles, is extensively used in medical diagnostics and represents the most widespread application of NMR. However, NMR is fundamentally limited in sensitivity and this has always restricted its applicability. Hyperpolarization techniques such as dynamic nuclear polarization (DNP) have become a major field of research and development because they hold the promise of increasing the sensitivity of NMR by several orders of magnitude. Such sensitivity enhancements could significantly broaden NMR applications, combining its unique structural information with much higher sensitivity. Unfortunately, there is no single implementation of DNP that would be suitable for a broader range of typical NMR applications. Experimental conditions often circumscribe areas of possible applications. Nevertheless, recent developments point towards experimental protocols providing solutions for specific applications of NMR. This review summarizes the concepts behind DNP in the light of recent developments and potential applications.

Keywords DNP • Dynamic Nuclear Polarization • Hyperpolarization • NMR

Dedicated to Heinz Rüterjans on the occasion of his 75th birthday.

U.L. Günther (✉)
HWB-NMR, School of Cancer Sciences, University of Birmingham, Vincent Drive, Edgbaston, Birmingham B15 2TT, UK
e-mail: u.l.gunther@bham.ac.uk

Contents

Abbreviations

CE	Cross effect
CIDNP	Chemically induced dynamic nuclear polarization
D-DNP	Dissolution DNP
DNP	Dynamic nuclear polarization
MRI	Magnetic resonance imaging
NMR	Nuclear magnetic resonance
O-DNP	Overhauser-DNP
OE	Overhauser effect
PHIP	Parahydrogen-induced polarization
QRP	Quantum rotor polarization
SE	Solid effect
SS-DNP	Solid-state DNP
TM	Thermal mixing

1 Introduction

1.1 Polarization

The widespread success of NMR spectroscopy arises from its vast information content and the predictability of spectra based on the solid theoretical framework underlying NMR [1]. However, NMR remains limited in sensitivity, as a direct consequence of the small energy difference between nuclear spin states. For $i = 1/2$ spins such as protons, ^{13}C, ^{15}N, ^{31}P, ^{19}F and other nuclei, the energy difference that determines the population of the two spin states α and β is $\Delta E = \gamma\hbar B_0$, where

γ is the gyromagnetic ratio and B_0 the magnetic field. This energy difference determines the population of spin states. Polarization is commonly defined as [2]:

$$P = \frac{n_\beta - n_\alpha}{n_\beta + n_\alpha} = \tanh \frac{\gamma \hbar B_0}{2kT}. \quad (1)$$

Equation (1) highlights the limitations of NMR: At the magnetic field of the largest NMR magnets available today, with $B_0 \approx 23.5$ T, equivalent to 1,000 MHz of proton frequency, the polarization is still only ≈80 ppm (0.008%). According to (1), polarizations are much higher at lower temperatures. At a temperature of 1 K the polarization of protons reaches 2.4% for 23 T, indicating a possible avenue for reaching higher polarizations. However, owing to its much larger gyromagnetic moment, unpaired electrons reach high polarizations at modest field strengths and temperatures below 10 K. DNP builds on the transfer of this high spin polarization from electrons to nuclear spins.

1.2 Hyperpolarization Methods

Besides DNP, there are several alternative routes towards increasing polarizations, building on a range of different experimental concepts. An obvious way towards larger polarizations lies in brute force methods. In principle, increasing the magnetic field falls into this category. This has been the method of choice in the past two to three decades, with the largest superconducting high-resolution NMR magnets now at 23.5 T. However, magnetic fields required to reach polarizations of 10–20% are far outside the window of what is currently feasible with superconducting magnets.

Brute force by lowering temperatures is a reasonable alternative for polarizing nuclear spins. At 0.1 K the polarization of protons reaches 23%. In practice, this is of course limited by the difficulty of reaching such low temperatures. Moreover, spectra of frozen samples need to be recorded with low-temperature magic angle spinning (MAS), which is technically enormously challenging at such temperatures. Alternatively, samples may be heated to ambient temperatures after polarization to acquire a high-resolution spectrum in the liquid state, a procedure that may well be feasible. There are, however, serious limitations to low temperature Boltzmann polarization, as the processes driving the polarization build-up will become very slow.

In another class of experiments, hyperpolarized states are generated by spin-sensitive chemical reactions. These include para-hydrogen-induced polarization (PHIP) [3–5] and chemically induced dynamic nuclear polarization (CIDNP) [6–8]. The latter involves non-equilibrium nuclear spin state populations that are produced in chemical reactions that proceed through radical pair intermediates. CIDNP's applicability has been focused towards the study of chemical reactions and the detection of surface exposed residues in proteins [9], but has so far remained limited to specialized chemical systems.

PHIP arises from quantum mechanical spin rotational states in diatomic hydrogen. Employing chemical reactions, these states can be used to generate molecules with enormously high spin polarizations. This is usually achieved by a hydrogenation reaction using para-hydrogen on an unsaturated organic molecule. The advantage of PHIP is that a theoretical 100% of ^{1}H polarization is feasible. PHIP has experienced a significant revival following recent developments by Duckett using a reversible approach whereby polarization is transferred via a metal centre [10], thus eliminating the need for an unsaturated analogue of the substrate. This approach now considerably broadens the applicability of PHIP as a general polarization method in solution and represents a competitive approach to DNP-NMR.

Quantum rotor polarization (QRP) has been known for a long time following pioneering work by Haupt in the 1970s [11]. QRP arises from rotational hindrance in methyl groups, giving rise to quantum tunneling. Large and rapid temperature jumps change the population of rotational energy levels and this process is associated with spin changes through the influence of the exclusion principle (reviewed in [12]). QRP can be seen as an equivalent to PHIP for molecules with C3 symmetry. Recently, new applications based on QRP have been explored by several groups [13–16], showing applicability beyond the narrow range of molecules known to exhibit such effects, if temperatures below 4 K are used [13]. QRP is also relevant for some variants of DNP as there the two mechanisms develop simultaneously and can interfere constructively or destructively.

Polarization of spins can also be achieved by optical pumping with circularly polarized laser light, primarily polarizing alkali metals (typically Rb), followed by spin exchange to noble gases (e.g., Rb–Xe spin exchange) [3, 17, 18]. Although limited to some noble gases, optically hyperpolarized ^{129}Xe and ^{3}He (both spin 1/2 nuclei) in the gas phase can be used for magnetic resonance imaging (MRI), either by using Xe directly [19] or by employing a nuclear Overhauser effect (NOE) to alter the polarization of nuclear spins [20]. Moreover, hyperpolarization of ^{83}Kr with spin 9/2 via Rb has been demeonstrated by Meersman and coworkers using natural abundance Kr containing 11.5% of the ^{83}Kr isotope [178].

Another method of polarizing nuclear spins is dynamic nuclear polarization (DNP), whereby the comparably large electron spin polarization (see Fig. 1) is transferred to nuclear spins by saturating the electron resonance. DNP is almost as old as NMR spectroscopy, building on the aforementioned theoretical work by Overhauser [21] who predicted what is today known as the Overhauser effect (OE). DNP was soon after demonstrated experimentally by Carver and Slichter [22, 23]. The enhancement, ε, that can be obtained by DNP is determined by the gamma ratio γ_e/γ_n, which is 660 for protons and 2,625 for ^{13}C.

DNP in the solid state, via the solid effect (SE) can be traced back to Pound [24], Abragam [25] and Jeffries [26]. Borghini described DNP in a spin temperature model [27, 28], which led to the discovery of thermal mixing (TM) [29]. In 1963 [30, 31], the cross effect (CE) was described, and later verified by Hwang, Hill [32, 33] and Wollan [34, 35]. In the following years, major work was carried out by Wind and coworkers [36]. Hausser and Stehlik derived the framework for DNP in

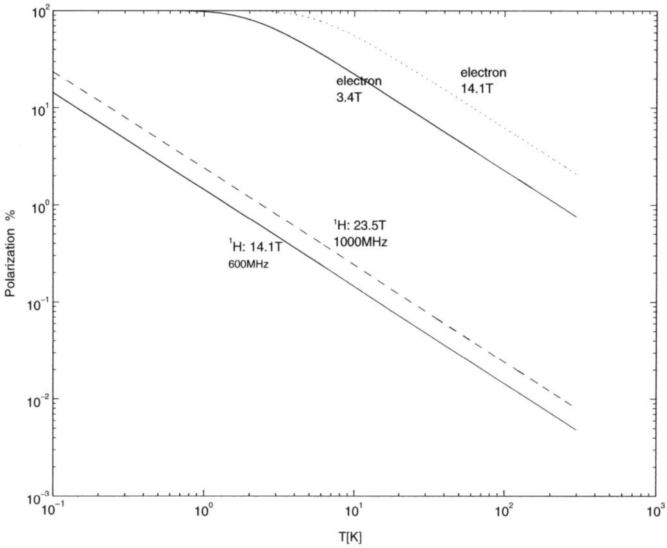

Fig. 1 Polarization levels as a function of temperature for ^{1}H at 14.1 T (600 MHz) and 23.5 T (1 GHz), and for electrons at 3.4 T and at 14.1 T, calculated using (1)

liquids [37], while Griffin [38] developed a major body of work leading to applications of DNP in solid state NMR using MAS.

The development of nuclear hyperpolarization has seen a significant revival in recent years, with major new developments in various hyperpolarization methods. This review focuses on DNP in liquids with the primary goal of providing an overview of the current state-of-the-art and the feasibility of applications arising from new developments. There are a number of reviews on DNP in liquids by Hausser and by Müller-Warmuth [37, 39], and a recent review by Bennati primarily focused on Overhauser DNP [40]. Early reviews include a chapter in a monograph by Abragam [1], and a review on the solid effect by Jeffries [2]. Abragam and Goldman published a fundamental review of theoretical concepts in 1972 [41]. More than 10 years later, Wind reviewed the developments of solid-sate DNP [36], summarizing the underlying theoretical concepts. A more recent review by Griffin provides a comprehensive overview of DNP principles and instrumentation [42].

The focus of this review is on DNP methods with applications in liquids. This includes dissolution DNP, which is of course strictly a solid-state polarization method although targeted towards NMR spectroscopy or imaging in the liquid state.

2 Overview Over DNP Mechanisms

Dynamic nuclear polarization describes a process whereby spin polarization is transferred from unpaired electrons to nuclear spins. This process exploits the much larger polarizations arising from electrons, which can be close to 100% at sufficiently low temperatures [see (1) and Fig. 1]. The choice of the transfer mechanism depends on the experimental conditions, whether the polarization is carried out in liquids or solids, and the choice and concentration of radicals. The transfer of polarization is initiated by irradiation at or close to the electron Larmor frequency, and the transfer involves simultaneous nuclear and electron spin transitions, during the irradiation or induced by relaxation.

The overall Hamiltonian of an electron-nuclear spin system in a magnetic field is given by:

$$H = \omega_e S_Z - \omega_n I_Z + H_{ee} + H_{en} + H_{nn}. \tag{2}$$

The first two terms describe the electron and nuclear Zeeman interactions, respectively, with $\omega_e = \gamma_e B_0$ and $\omega_n = \gamma_e B_0$, where γ_e and γ_n are the gyromagnetic ratios of electrons and nuclei, respectively, and B_0 represents the external magnetic field strength. S and I are used for electron and nuclear spin operators, respectively. For indices, this review always uses n for nucleus and e for electrons, i.e., $\gamma_S = \gamma_e$ and $\gamma_I = \gamma_n$.

H_{ee}, H_{en} and H_{nn} denote the spin–spin interactions between electrons, between electrons and nuclei, and between nuclei, respectively. H_{ee} has often been neglected for sufficiently dilute concentrations of electrons. It becomes, however, relevant for the solid state DNP mechanisms (SE, TM and CE); in the case of the CE a specific electron–electron interaction can be utilized to optimize the DNP enhancement. It is also possible that electron relaxation is important for polarizations carried out at very low temperatures.

Abragam and Goldman recognized that the dipolar interaction causing relaxation between nuclei determines the polarization time in the solid state. Nuclear relaxation is also relevant in solution, as will be shown in 2.1. Moreover, recent work by Vega has paid attention to nuclear spin diffusion associated with DNP, the key mechanism for transfering magnetization from molecules in contact with the radical to other nuclei in the solid state [43].

In solution, the term that seems most important for DNP is H_{en}, describing the electron–nuclear interactions, which are governed by the dipolar interaction H^D and the scalar Fermi contact interaction H^{Sc}, where:

$$H_{en}^D = \frac{\gamma_n \gamma_e \hbar}{r_{jk}^3} [I \cdot S - 3(I \cdot n)(S \cdot n)], \tag{3}$$

where r is the distance between the electron and the nucleus and $\boldsymbol{n} = \boldsymbol{r}/r$ is the unit vector in the direction joining centres of the two particles.

In high magnetic field, the non-secular part is often discarded leaving solely the secular part of (3):

$$H_{en}^D = \frac{\gamma_n \gamma_e \hbar}{r_{ij}^3} [I \cdot S - 3 I_z S_z]. \tag{4}$$

The Hamiltonian for the scalar Fermi contact can be written as:

$$H_{\mathrm{en}}^{\mathrm{Sc}} = I \cdot A \cdot S, \tag{5}$$

where A represents the scalar hyperfine coupling tensor.

All mechanisms of DNP share the common principle of transferring polarization from electrons to nuclei, following irradiation at or near the frequency of the ESR line of the electron. The DNP process depends essentially on the magnitude of the hyperfine and dipolar interactions. However, this process only determines the polarization of nuclei in close proximity to an electron, the core nuclei. In liquids, this polarization transfer is driven by relaxation, following excitation of electron transitions. In solids, transitions are directly driven by microwave irradiation.

For the DNP process to proceed from an initial polarization zone, it is essential that polarization be either transported to other nuclei via spin diffusion, as is the case in the solid state. Alternatively, in liquids unpolarized nuclei become available by translational diffusion. These two options are responsible for largely different polarization processes in (insulating) solids as compared to liquids (and metals). In liquids, this process is fundamentally different, and rotational and translational diffusion processes govern the polarization process, as originally described by Hausser et al. [37], Müller-Warmuth et al. [39] and others. Relaxation processes are driven by the time dependence of the overall Hamiltonian, arising form variations in the vector $\boldsymbol{r}$ and the hyperfine coupling constants, causing relaxation transitions between the spin states.

The choice of the polarization mechanism depends entirely on the experimental conditions. Overhauser DNP is the only DNP mechanism known to work in liquids. The specific mechanism in the solid state depends primarily on the experimental conditions, i.e., choice and concentration of the radical. Table 1 summarizes the DNP mechanisms and the conditions required for these mechanisms. The key mechanisms will be reviewed in the following paragraphs.

DNP can also be classified by the experimental arrangement that is used to carry out the experiment. In this sense three technical implementations of DNP prevail:

1. Overhauser DNP (O-DNP), driven by the Overhauser effect (OE), where the entire process is carried out in the liquid state.
2. Solid-state DNP (SS-DNP) at low temperatures (typically 90 K), driven by the cross effect (CE), using high-power microwave sources in conjunction with MAS for solid-state NMR.
3. Dissolution DNP or ex situ DNP (D-DNP), typically carried out at temperatures <1.5 K, driven by thermal mixing (TM) or the solid effect (SE). The polarized liquid sample is generated by sample dissolution and is transfer to an analytical NMR magnet to acquire a spectrum.

Significant progress in DNP arises from improved instrumentation, mainly from high power microwave sources that have become available in recent years. These developments were much influenced by Griffin's pioneering work using gyrotron

Table 1 Common DNP mechanisms

Underlying mechanism	OE	SE	CE	TM
Common application	O-DNP (liquids)	D-DNP	SS-DNP	D-DNP
Sample type	Liquids; alkali metals; organic conductors	Glassy samples in the solid state	Glassy samples in the solid state	Glassy samples in the solid state
Conditions	H_{en} time-dependent $\approx\omega_e^{-1}$; polarization transfer via relaxation, depending on scalar and dipolar interaction.	H_{en}^{Sc} causes mixing of states ⇒ forbidden transition, slow build-up Low concentration of radicals, no dipolar coupling between radicals	Inhomogenous broadening of the ESR signal, large g-anisotropy, dipolar coupling between radicals. Optimal with biradicals satisfying: $\omega_n = \omega_{eS1} - \omega_{eS2}$	Homogenous broadening[a] of the ESR signal, small g-anisotropy High concentration of radicals
		$\Delta\omega_{e1/2} \ll \omega_n$ Unresolved SE for $\omega_{e1/2} \approx \omega_n$	$\Delta > \omega_n > \omega_{e1/2}$ ⇒ resolved hyperfine lines in ESR spectrum	$\omega_{e1/2} > \omega_n$
Frequency dependence	$\omega_e\tau_c < 1$	~B_0^{-2}	B_0^{-1}	$\omega_{e1/2}$
Microwave irradiation frequency for maximal enhancement	$\omega = \omega_e$	$\omega = \omega_e \pm \omega_n$ Maxima in microwave sweep separated by $2\omega_n$	ω_{max} at polarization maximum, depends on EPR line shape	ω_{max} at polarization maximum, depends on EPR line shape $\omega \approx \omega_e \pm \omega_{e1/2}$

ω_e electron Larmor frequency, ω_n nuclear Larmor frequency, $\omega_{e1/2}$ homogenous half width of the ESR signal, Δ inhomogenous breadth of the EPR spectrum, τ_c electron rotational correlation time.
[a]The expressions homogenous and inhomogenous broadening are commonly used in EPR spectroscopy, but are less familiar to NMR spectroscopists. Inhomogeneous broadening of the ESR spectrum describes a signal composed of a large number of narrower spectra, arising from different hyperfine couplings, each shifted with respect to each other. In the case of homogeneous broadening, the linewidth of the ESR spectrum is determined by the electron relaxation times

sources for SS-DNP [42]. Another important technological development is that of D-DNP by Ardenkjær-Larsen [44], which yields by far the largest overall enhancements and has major applications in MRI. Moreover, recent work on O-DNP shows an avenue towards applications of DNP in solutions. Implementations of O-DNP come in different flavors, either carrying out the polarization at the same field strength as the NMR measurement or polarizing at lower field and transporting the sample to higher field for the NMR measurement.

2.1 Overhauser DNP

The OE was first described theoretically in a 1953 landmark paper by Albert Overhauser, who predicted for metals that "if the electron spin resonance of the conducting electrons is saturated, the nuclei will be polarized to the same degree they would if their gyromagnetic ration were that of the electron spin"[21]. In the same year, this theoretical concept was verified by Carver and Slichter for metallic lithium dispersed in oil, resulting in an enhancement factor of 100-fold [22]. Figure 2 shows the first DNP spectrum from a 1953 publication by Carver and Slichter [22]. Later the same authors demonstrated the DNP effect for ^{23}Na and ^{1}H in sodium ammonia solutions [23]. The history of these early DNP experiments was recently reviewed by Slichter [45].

Although O-DNP was originally discovered in alkali metals, its main application today is in DNP in solution, predominantly for ^{1}H and for ^{13}C NMR, although many other spin 1/2 nuclei can be polarized. Earlier applications also included ^{1}H and ^{13}C DNP of organic conductors [46–48].

The O-DNP experiment requires irradiation at the electron Larmor frequency to saturate the electron transition. The enhancement arises from subsequent relaxation processes involving simultaneous reversals of I and S in opposite directions (flip-flop transitions, W_0), or in the same direction (flip-flip transitions, W_2). This is depicted by the energy level diagram in Fig. 3. Hausser and Stehlik introduced a phenomenological description using rate equations [37], based on Solomon's treatment of the OE. According to the Solomon equations [49], the rate equation for the expectation value of the nuclear polarization $<I_z>$ can be written as:

$$\frac{\mathrm{d}\langle I_z\rangle}{\mathrm{d}t} = -\rho(\langle I_z\rangle - I_0) - \sigma(\langle S_z\rangle - S_0), \tag{6}$$

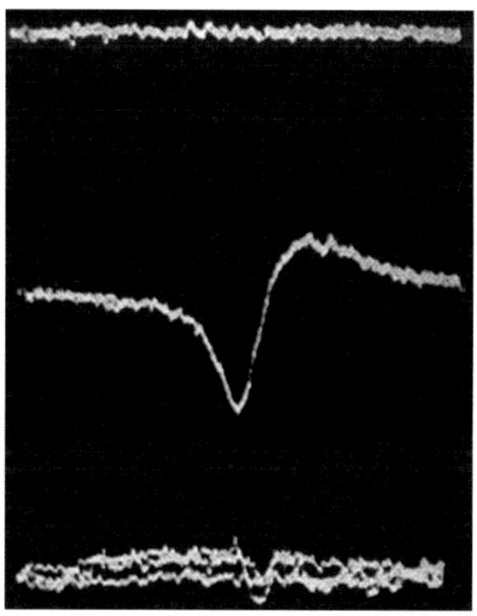

Fig. 2 Original DNP experiment performed by Carver and Slichter [22]. *Top*: The noisy ^{7}Li resonance. *Middle*: The ^{7}Li resonance enhanced by electron saturation. *Bottom*: Proton resonance in glycerin sample. (Copyright 1953 by The American Physical Society)

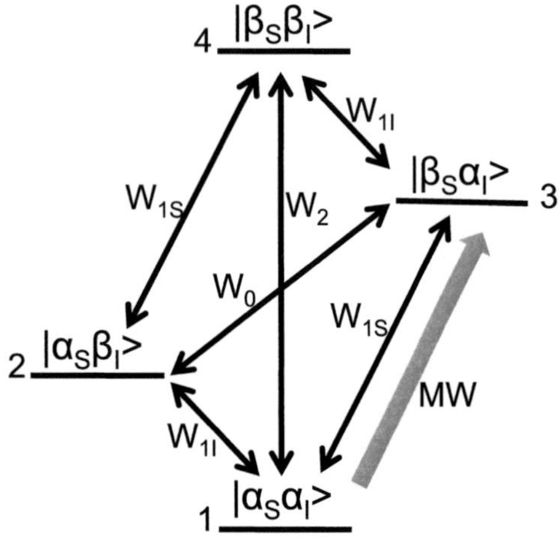

Fig. 3 Energy level diagram for a two-spin system; S represents the electron and I a spin 1/2 nucleus. Transitions are labelled with the respective transition probabilities: W_{1S} and W_{1I} are the transition probabilities for the electron S and for the nucleus I, respectively. W_0 and W_2 represent the zero and double quantum transition probabilities, respectively

where I_0 and S_0 represent the values of $<I_z>$ and $<S_z>$ in thermal equilibrium. Hausser introduced three key parameters, the coupling factor ξ, the leakage factor f and the saturation factor s [37]:

$$\xi = \frac{\sigma}{\rho} = \frac{W_2 - W_0}{W_0 + 2W_1 + W_2}, \tag{7a}$$

where σ is the cross-relaxation and ρ the auto-relaxation rate,

$$f = \frac{\rho}{\rho + W_{1I}^0}, \tag{7b}$$

$$s = \frac{S_0 - \langle S_z \rangle}{S_0}. \tag{7c}$$

The coupling factor describes the contribution of the transition probabilities W_i towards the maximum polarization and depends on the nature of the short range interactions that determine these transition probabilities.

The overall DNP enhancement has been defined as:

$$\epsilon = \frac{\langle I_z \rangle}{I_0} = 1 - \xi \cdot f \cdot s \frac{|\gamma_e|}{\gamma_n}. \tag{8}$$

It should be noted that the commonly used NOE factor is defined differently as $\frac{I_z - I_0}{I_z} = 1 - \epsilon = \xi \cdot f \cdot s \frac{|\gamma_e|}{\gamma_n}$. Some authors defined their DNP enhancement this way. If f and s both have a value of +1 the enhancement is $\epsilon = 1 - \xi \frac{|\gamma_e|}{\gamma_n}$, i.e., a sufficiently large positive enhancement reflects a negative coupling factor.

The leakage factor f measures the contribution of the electrons towards the relaxation of the nuclei. (W_{1I}^0 in (7b) is the longitudinal relaxation rate of the nucleus in the absence of a radical). It can be determined by relaxation measurements in the presence and absence of the radical. Values can range from 0 (no contribution from the radical) to 1 (relaxation controlled by the radicals). For nitroxide radicals it usually assumes a value close to 1 at radical concentrations of 10–20 mM.

The saturation factor *s* is a measure of the degree to which the electron transitions are saturated. When the electron transitions are completely saturated, *s* takes a value of 1. It depends on the microwave power and the relaxation of the electrons. The saturation factor becomes significantly more complicated in radicals with several hyperfine lines (two for ^{15}N, three for ^{14}N), such as nitroxide radicals, where various relaxation mechanisms need to be considered if only one line is saturated. Bates and Drozdoski proposed a theoretical model taking into account electron spin exchange transitions [50], which has recently been extended by Armstrong [51] to account for the relaxation effect arising from ^{14}N (and ^{15}N) nuclei. Recently Türke and Bennati [179] proposed to measure saturation factors using pulsed electron–electron double resonance (ELDOR) experiments measuring the saturation level of a hyperfine line when pumping a coupled line of the nitroxide spectrum. For ^{15}N-^{2}H-labelled TEMPONE (4-oxo-2,2,6,6-tetramethylpiperidine-*N*-oxyl) radical, this study demonstrates a strong dependence of *s* on the concentration of the radical, owing to a predominant Heisenberg exchange mechanism at concentrations $\geq$5 mM. Almost complete saturation can be achieved in water by irradiating one electron paramagnetic resonance (EPR) transition. According to calculations by Armstrong et al., this may not be the case for ^{14}N-TEMPONE [51] (radicals are described in Sect. 3.4).

Hausser used a semi-classical approach to determine the dependence of σ and ρ on the spectral densities of molecular motions for different relaxation mechanisms (S for scalar, D for dipolar) [49]:

$$\rho_{\mathrm{S}} = -\sigma_{\mathrm{S}} = \left[\left(\frac{A^2}{2}\right)J_{\mathrm{S2}}(\omega_{\mathrm{n}} - \omega_{\mathrm{e}}) + \beta J_{\mathrm{S1}}(\omega_{\mathrm{n}})\right], \tag{9}$$

$$\rho_{\mathrm{D}} = \frac{\gamma_{\mathrm{n}}^2\gamma_{\mathrm{e}}^2 h^2}{10 r_{\mathrm{en}}^6}\left[(6J_{\mathrm{D}}(\omega_{\mathrm{n}} + \omega_{\mathrm{e}}) + J_{\mathrm{D}}(\omega_{\mathrm{n}} - \omega_{\mathrm{e}}) + 3J_{\mathrm{D}}(\omega_{\mathrm{n}})\right], \tag{10}$$

$$\sigma_{\mathrm{D}} = \left(\frac{\gamma_{\mathrm{n}}^2\gamma_{\mathrm{e}}^2 h^2}{10 r_{\mathrm{en}}^6}\right)\left[6J_{\mathrm{D}}(\omega_{\mathrm{n}} + \omega_{\mathrm{e}}) - J_{\mathrm{D}}(\omega_{\mathrm{n}} - \omega_{\mathrm{e}})\right], \tag{11}$$

with Lorentz-shaped scalar and dipolar spectral density functions for rotational diffusion [52]:

$$J_{\mathrm{S1,2}}(\omega) = \tau_{\mathrm{S1,2}}/(1 + \omega^2\tau_{\mathrm{S1,2}}^2); \quad J_{\mathrm{D}}(\omega) = \tau_{\mathrm{D}}/(1 + \omega^2\tau_{\mathrm{D}}^2). \tag{12}$$

The correlation times of the scalar and dipolar couplings are:

$$\tau_{\mathrm{S1,2}} = (1/\tau_{1,2} + 1/\tau_{\mathrm{e}})^{-1} \quad \text{and} \quad \tau_{\mathrm{D}} = (1/\tau_{\mathrm{e}} + 1/\tau_{\mathrm{r}})^{-1}. \tag{13}$$

In these equations, Hausser introduced the last term $\beta J_{S1}(\omega_n)$ in (9) to account for a time dependence of scalar relaxation by spin exchange, in which case electronic relaxation transitions induce pure nuclear relaxation transitions. β takes a value of 1 for $\tau_e \gg \tau_1, \tau_2$ and 0 for $\tau_e \ll \tau_1, \tau_2$. From these spin density functions and their contribution towards ρ and σ, the dependence of the coupling factor ξ can be calculated. Such curves are shown in Fig. 4.

These models are still limited by the restriction to rotational tumbling of both the electron and the nucleus. Müller-Warmuth and coworkers developed combined rotational and translational diffusion models (reviewed in [39]) for dipolar and scalar interactions, assuming independent diffusion of the molecules. They also developed a pulse diffusion model assuming occasional collisions between molecules, described via a Poisson process [39]. Later Hwang et al. [53] used force-free pair correlation functions to account for translational diffusion and ionic interactions for dipolar interactions, leading to the now commonly used equation:

$$J_{\text{trans}} = \frac{1 + \frac{5z}{8} + \frac{z^2}{8}}{1 + z + \frac{z^2}{2} + \frac{z^3}{6} + \frac{4z^4}{81} + \frac{z^5}{81} + \frac{z^6}{648}}, \quad \text{with } z = \sqrt{2\omega\tau_{\text{TD}}}, \tag{14}$$

where τ_{TD} is the correlation time for the translational diffusion constant, which has often been related to the closest distance between molecules d and to the diffusion coefficients D_I and D_S by $\tau_{TD} = d^2/(D_I + D_S)$. τ_{TD} and d can be determined by fast field-cycling relaxometry (NMR dispersion, NMRD) [54, 55]. Translational

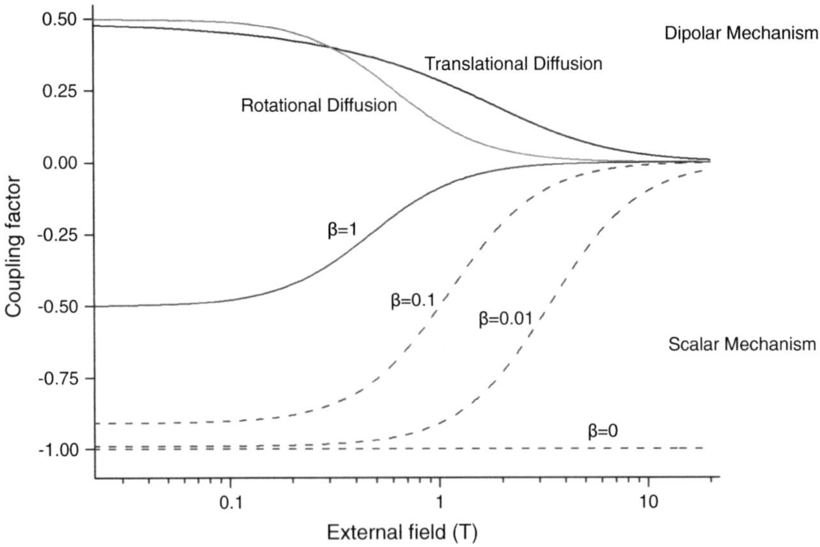

Fig. 4 Dependence of the coupling factor ξ derived from (9)–(13) on the magnetic field (and therefore on the Larmor frequency) for scalar and dipolar relaxation using spectral densities for rotational and translational diffusion using a correlation time of 20 ps (Reproduced from [40] with the permission of the Royal Society of Chemistry)

diffusion yields somewhat flatter curves than those for rotational diffusion, despite an overall similar behaviour (Fig. 4).

Overall, this treatment predicts that the scalar term for rotational diffusion is independent of double quantum relaxation (W_2) and depends only on zero quantum relaxation. The coupling factor ξ can assume values between 0.5 and -1.0 for pure dipolar and pure scalar relaxation, respectively. Moreover, the curves in Fig. 4 clearly show the expected field dependence with low values for the coupling factor, and therefore low enhancements for high magnetic fields.

The experimental behaviour of the saturation factor at low field (0.34 T) has been summarized by Müller-Warmuth [39] considering data up to 1983. The experimental curves for different nuclei and a series of radicals (Figs. 7, 8 and 10 in [39]) show a substantially more complex behaviour than the theoretical curves shown in Fig. 4 with mixed scalar and dipolar contributions. The scalar and dipolar contributions depend strongly on the nature of the sample, in fact on the specific nucleus, the solvent, and the choice of radical.

The relaxation mechanism encountered is therefore not predictable. As a general rule, protons have almost always shown a coupling factor driven by dipolar relaxation. Known exceptions are aromatic protons, for which Müller-Warmuth reported a weak scalar contact coupling, and acid protons of trifluroacetic acid [39]. For ^{13}C, positive and negative enhancements were reported. A more complex relaxation mechanism must be considered, particularly when the carbon is bound to a proton, in which case we have a three-spin system and possibly an additional ^{1}H/^{13}C-NOE. Most carbons seem to favour a dipolar mechanism (negative enhancement, positive ξ); recent examples include small molecules such as urea, acetone (CO and CH_3), pyruvate and methanol [56, 57], whereas *N*, *N*-dimethylformamide [57], chloroform [56] and other chlorinated sp3 carbons [39] showed a scalar coupling mechanism, as indicated by a positive enhancement and a negative ξ. Loening reported scalar relaxation for ^{31}P in triphenylphosphine, ^{13}C in carbontetrachloride, ^{15}N in aniline (opposite sign of the enhancement as $\gamma_N^{15} < 0$) and for ^{19}F in hexafluorobenzene. ^{31}P is an interesting case for studying the underlying effects, as trivalent phosphorus seems to favour a scalar relaxation mechanism, probably arising from lone pair interactions with the electron [39].

The temperature dependence of O-DNP has been the subject of recent discussions as it influences correlation times and the overall enhancement factor. The temperature dependence has been studied by NMRD measurements of water [54] and by molecular dynamics simulations [58], and more recently by measuring temperatures from the chemical shift change of the water resonance [59]. Temperature affects the saturation factor s by broadening the EPR lines of the TEMPONE nitroxide radical and, more importantly, the coupling factor ξ by the change of correlation times. These measurements showed a linear increase of the enhancement with temperature [59], in good qualitative agreement with the values of ξ from molecular dynamics simulations [58], but showed a significantly higher increase with temperature than previously expected. Bennati determined ξ using correlation times from NMRD measurements and found a somewhat more rapid variation with temperature than those reported by Kryukov [59]. Overall, significant heating despite using small capillary samples (typically of nanolitre volumes) represents a major obstacle in a systematic analysis of the polarization parameters, which may have obscured previous results.

2.2 DNP via the Solid Effect

Soon after the discovery of DNP via the OE, Jeffries realized that "the saturation of so-called forbidden transitions will produce a comparable nuclear polarization directly in the sense that the applied radiofrequency field itself flips the nuclei" in solids [26]. He mentions that this had previously been observed by Pound [24] and theoretically considered by Abragam [25], who coined the name "l'effet solide." Jeffries later described the SE theoretically [60], and reviewed the early work of DNP in 1964 [2].

The SE is observed at low temperatures in solids for radicals with narrow EPR lines (typically trityl radicals), where the EPR linewidth is much smaller than the nuclear Larmor frequency ($\omega_{e1/2} \ll \omega_N$). As pointed out by Jeffries, it arises from driving the forbidden W_0 and W_2 as illustrated in Fig. 5.

DNP in the solid state follows a different mechanism compared to DNP in liquids, in particular at low temperatures, where the spatial parts of H_{en}^{D} and H_{en}^{Sc} become time-independent. Therefore, the enhancement is not driven by changes in polarizations arising from W_2 and W_0 relaxation, as is the case for O-DNP, but rather by driving these forbidden flip-flop (W_0) and flip-flip (W_2) transitions using microwave irradiation at the corresponding resonance frequencies $\omega_e + \omega_N$ or $\omega_e - \omega_N$, as shown in Fig. 6.

For this it is useful to expand the Hamiltonian (2) to include a term for the interaction of the electrons with the microwave field:

$$H = \omega_{0e} S_Z - \omega_{0n} \sum_i I_Z^i + 2\omega_{1e} S_Z \cos \omega t + \sum_i \mathbf{S} \cdot \mathbf{A}^i \cdot \mathbf{I}^i + H_{nn}. \quad (15)$$

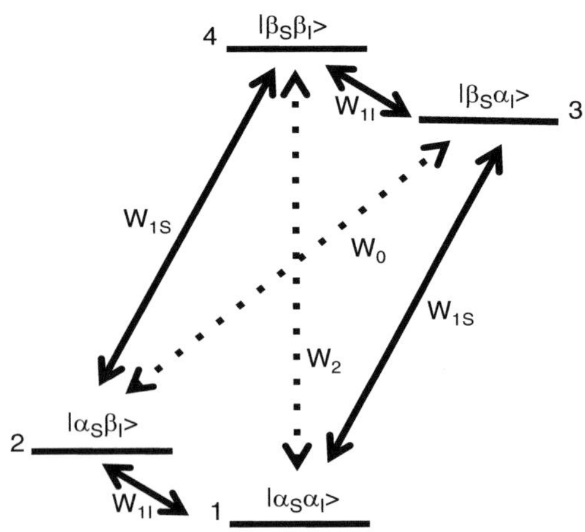

Fig. 5 Energy level diagram for the SE, which is driven by forbidden single and double quantum transitions (W_0 and W_2, *dotted lines*)

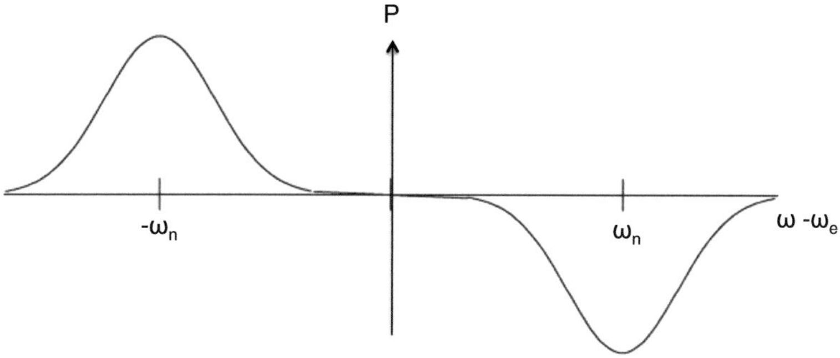

Fig. 6 Nuclear polarization by the SE as a function of the microwave frequency

The first and second term describe electron and nuclear Zeeman interactions, of the electron spin and the nuclear spins with the applied magnetic field, respectively, the third term the interaction of the electrons with the microwave field, and the fourth the hyperfine interaction between the nuclear spins. Here $\omega_{0e} = \gamma_e B_0$ is the electron Larmor frequency, with the applied magnetic field B_0 and the gyromagnetic ratio of the electron γ_e, and $\omega_{0n} = \gamma_n B_0$ the nuclear Larmor frequency with the nuclear gyromagnetic ratio γ_n. For the interaction of the electron spin with the microwave field B_1 we use $\omega_{1e} = \gamma_e B_1$. $\mathbf{A}^i$ represents the hyperfine tensor for the interaction between nuclei and electron spins. The term H_{nn} describing dipolar interactions between nuclear spins has been neglected in theoretical descriptions of the SE by Wenckebach [61, 62] and by Jeschke [63], but has been the subject of a recent theoretical treatment of coupled solid state DNP and spin diffusion using the density matrix formalism by Hovav and Vega [43].

The SE depends on non-secular terms of the electron-nuclear part of the Hamiltonian, leading to the mixing of states under the influence of microwave irradiation. This can easily be recognized by writing the Hamiltonian in a frame that points into the direction of the effective microwave field in which the z′ axis is defined by effective microwave frequency:

$$\omega_{\mathrm{eff}} = \sqrt{(\omega_{0e} - \omega)^2 + \omega_{1e}^2} - \sqrt{\Delta^2 - \omega^2}. \tag{16}$$

Here, ω is the microwave frequency, and $\omega_{0e} - \omega$ represents the detuning of the microwave field with respect to the resonance frequency of the electron spin.

Polarization under the SE occurs when the Hartmann–Hahn condition $\omega_{\mathrm{eff}} = \omega_n$ is fulfilled. The Hartmann–Hahn condition has two solutions $\Delta_{1,2} = \pm\sqrt{\omega_n^2 - \omega_{1e}^2}$, which yield maximum polarizations of opposite sign at the microwave frequencies $\omega_\pm = \omega_{0e} \pm \omega_n \cos(\Theta_{1,2})$, where the tilt angle of the effective field $\Theta_{1,2} = \mp\frac{\sqrt{\omega_n^2 - \omega_{1e}^2}}{\omega_n}$. For sufficiently large B_{1e} field strengths, Θ adopts small values and $\cos(\Theta) \rightarrow 1$, in which case $\omega_{1,2} \approx \omega_{0e} \pm \omega_n$.

The new Hamiltonian, ignoring the H_{nn} term for nuclear spin diffusion, and after several transformations into the tilted frame and averaged over a period of $2\pi\omega_{0n}^{-1} \approx 2\pi\omega_{\text{eff}}^{-1}$, thus satisfying the Hartmann–Hahn condition, we get:

$$H_{\text{tilt,av}} = Z + V_{zz}\cos\Theta + V_{\pm}\sin\Theta, \tag{17}$$

where

$$Z = \omega_{\text{eff}} S_{z'} + \omega_{0n}\sum_i I_{z'}^i,$$

$$V_{ZZ} = \sum_i A_{zz}^i S_{z'} I_z^i$$

and

$$V_{\pm} = \tfrac{1}{4}\sum_i \left(A_{z+}^i S_+ I_-^i + A_{z-}^i S_- I_+^i\right).$$

The last term represents the flip-flop terms causing polarization transfer. Here, $S_{\pm} = S_{x'} \pm iS_{y'}$ and $I_{\pm} = I_{x'} \pm iI_{y'}$ are the step operators, and $A_{z\pm} = A_{zx} \pm iA_{zx}$ represents the tensor elements of the hyperfine coupling. The detailed transformation can be found in [61, 62]. Wenckebach developed this approach to describe coherent polarization transfer in the rotating frame; this is required for the nuclear orientation via electron spin locking (NOVEL) pulsed DNP experiment [64] in which a spin lock is applied to the electrons.

The transition probability for the forbidden transitions $W_{\pm}$ are given by:

$$W_{\pm} = 2\left|q_{ij}\right|^2 \pi\omega_e^2 f(\omega_e \pm \omega_n), \tag{18}$$

where q_{ij} are the coefficients that describe the degree of Eigenstate mixing $(q_{ij}|+>_i| + >_i)$, which is given by:

$$q_{ij} = -\frac{3}{4}\frac{\gamma_e\gamma_n\hbar}{\omega_n}\cdot\frac{1}{r_{ij}^3}\cdot\sin\theta_{ij}\cos\theta_{ij}e^{-i\phi_{ij}}. \tag{19}$$

$f(\omega_e \pm \omega_n)$ represents the normalized EPR lineshape function. In this equation, W_2 is proportional to q_{ij}^2 and therefore to ω_n^{-2} and B_0^{-2} (in contrast to CE and TM, which scale with B_0^{-1}) [36, 65]. This is thought to limit the use of the SE at higher field strengths.

If the ESR line is sufficiently narrow and the nuclear Larmor frequency sufficiently large $(\omega_n \gg \omega_{e1/2})$, polarization transfer occurs in two completely separated frequency regions, as shown in Fig. 6. In the case of the "differential solid effect" these lines are not as well resolved; consequently, the SE becomes inefficient, owing to simultaneous transitions of opposite sign that may cancel each other. This becomes relevant for low magnetic field strengths and for low γ nuclei. In practice, it may be hard to distinguish this situation from the thermal mixing DNP mechanism.

By solving rate equations similar to those used to describe the OE one can readily show under which conditions complete polarization becomes possible. Wind [36] derived the SE enhancement factor as:

$$\epsilon_{SE} = 1 + \frac{W_2}{W_2 + W_{1I}} \cdot \left(\frac{\gamma_e}{\gamma_n} - 1 \right), \tag{20}$$

under the assumption that W_{1e} dominates the cross-relaxation rates W_2, W_0 and W_{1n} ($W_{1e} \gg W_2$,W_0,W_n). This enhancement becomes γ_e/γ_n if cross-relaxation dominates the nuclear relaxation rate ($W_2 \gg W_n$). The total enhancement is scaled by $\cos(\theta)$, where θ is the tilt angle of the effective field versus the static magnetic field, although θ will be small for sufficiently large magnetic field strengths.

Polarizations in the solid state work best in solid glass-forming solvents, such as glycerol, water, methanol, DMSO and certain mixtures of those. Deuteration considerably improves the polarization because it removes protons as the most important relaxation mechanism. It has also been shown that ^{13}C-enrichment of the glassy matrix with compounds like ^{13}C-acetone or ^{13}C-DMSO improves the polarization process, in some cases by enhancing the achievable polarization [66], for ^{13}C-DMSO and pyruvate by increasing the polarization rate [67]. These experiments highlight the importance of the spin diffusion barrier surrounding the paramagnetic centre [68], which has been the issue of recent theoretical considerations [43, 69]. Moreover, for polarizations at low temperature an additional relaxation mechanism arising from quantum rotor transitions in methyl groups must be considered [13, 70]. For polarizations of some substances this may dominate to a degree where the polarization arising from the CE may be completely quenched, although this effect can be eliminated by choosing the microwave frequency at $\omega_e + \omega_n$, where quantum rotor effects and DNP enhancements interfere constructively [70].

2.3 DNP via the Cross Effect and Thermal Mixing

The CE was discovered when Hwang examined Ley's radical (a bis-phenyl aroxyl radical) in polystyrene for increasing concentrations of the radical (0.25–5%) and found that the positive and negative DNP enhancement peaks move closer to each other for increasing concentrations of the same radical, as depicted in Fig. 7b [32, 33]. The underlying effect had previously been proposed by Abragam and Borghini [71] arising from spin–spin interactions between the electrons. A theoretical treatment by Wollan [35] also mentions an earlier discovery by Kessenikh [30, 31]. Despite its importance for SS-NMR, there is no comparable theoretical basis for CE as there is for the SE, although the basic requirements of the CE are reasonably well understood.

Both, the CE and TM require that the inhomogeneous EPR linewidth Δ is larger than the nuclear Larmor frequency ($\Delta > \omega_n$), while the homogenous width must be

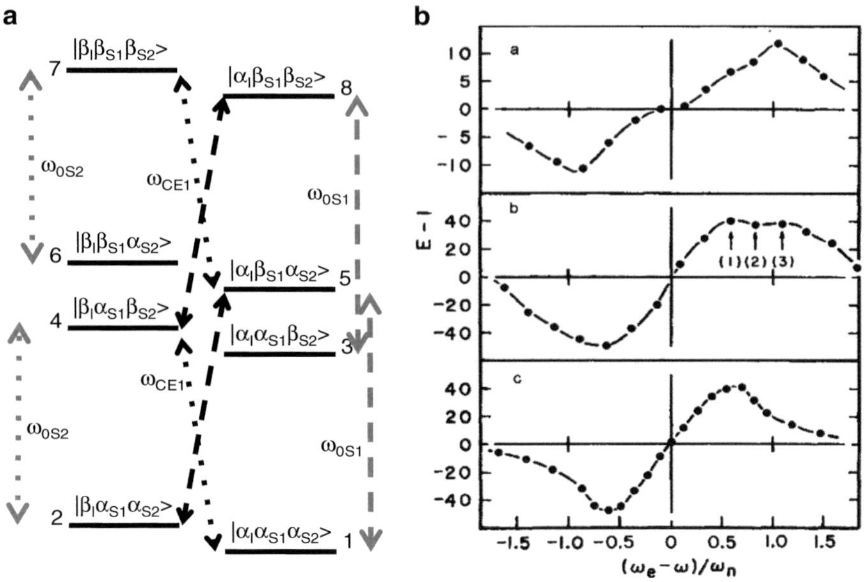

Fig. 7 (**a**) Energy level diagram for the CE, which requires a three-spin system with two electrons and one nucleus. Saturation of one of the EPR transitions for one of the dipolar coupled electrons (ω_{CE1} or ω_{CE2}) yields the enhancement for the nuclear transitions (ω_{0S1} for ω_{CE1} causing a negative enhancement, ω_{0S2} for ω_{CE2} causing a positive enhancement). (**b**) Polarization sweeps for protons obtained by Hwang et al. for increasing concentrations of Ley's radical in polystyrene. (0.25, 1.5 and 5%). The positive and negative DNP enhancement peaks move closer to each other for increasing concentrations of radical [32]. (Copyright 1967 by The American Physical Society)

smaller than the nuclear Larmor frequency ($\omega_{e1/2} < \omega_n$). For homogenously broadened EPR lines TM becomes active, for inhomogenously broadened lines CE is encountered. The latter is the case at higher fields and for larger g-tensors. The CE and TM are based on allowed transitions (Fig. 7a) and builds up polarization much faster than the SE.

Both effects require two or more electrons in close proximity. For CE, two electrons flip simultaneously with the nuclear spin via $\left|\beta_I\beta_{S1}\alpha_{S1}\right> \left|\alpha_I\alpha_{S1}\beta_{S1}\right>$ and $\left|\alpha_I\beta_{S1}\beta_{S2}\right> \left|\beta_I\alpha_{S1}\alpha_{S2}\right>$ transitions. This is achieved by irradiating at the frequency of one of the two radicals; as a consequence, both radicals and the proton flip. The matching condition for these transitions is $|\omega_{e1} - \omega_{e2}| \approx \omega_n$. This is the case when the levels 4 and 5 become degenerate, which can be approximated at high radical concentrations. Griffin and coworkers developed the concept of biradicals (such as TOTAPOL) for which the two electrons fulfill the matching condition. As the linewidth of EPR spectra scales with B_0, the effect scales with B_0^{-1} [42].

TM is similar to the CE but has been developed using a different theoretical model, specifically the concept of spin temperature based on the Provotov theory [1, 72], in which spin systems are treated as thermodynamic ensembles, characterized by a spin temperature. This concept is suitable for describing a system with many electrons, where the Hamiltonian becomes too complex to determine individual Eigenlevels. This theory has recently been reviewed by Goertz [73], and has been applied to DNP spectra of solid deuterated butanol [74].

Qualitatively, TM can be interpreted as a transition from the CE towards an electron spin ensemble, arising from manifolds of states. TM requires high radical concentrations causing a homogenously broadened spectrum arising from strong electron interactions.

2.4 Pulsed DNP Methods

Pulsed DNP methods were recently reviewed [42]. Without doubt they offer significant advantages over classical DNP approaches. They do not scale as unfavourably with increasing field strengths, and have the potential to overcome some of the limitations encountered with SE, TM and CE. Pulsed methods include rotating frame DNP [65, 75], the integrated SE [76, 77], NOVEL [62, 64], the dressed SE [78] and a recently described method termed PONSEE (polarization of nuclear spins enhanced by ENDOR) [79]. NOVEL is a coherent method, which might have the potential to enable much faster polarizations, overcoming some of the limitations of the SE. However this requires irradiation at high power to fulfill a Hartmann–Hahn condition, which is not easily possible at higher field strengths with currently available microwave technology. Nevertheless, pulsed methods may have significant advantages for future experiments.

3 Experimental Implementations and Applications

The choice of experimental settings has largely been dictated by the constraints predicted by theoretical treatments, and by experimental limitations. SS-DNP is quickly becoming an important add-on to solid-state MAS-NMR spectrometers, building on Griffin’s work using high power gyrotron microwave sources [80, 81]. These microwave sources open new avenues towards applications of DNP in biological systems. SS-DNP will not be covered in great detail because it is described in several excellent reviews by Griffin [42, 180]. The most important highlights of this work will, however, be summarized briefly in Sect. 3.1.

For DNP in solution, high power microwave irradiation is not necessarily the best possible concept owing to the heating effects in many solvents. Nevertheless, Prisner and coworkers have presented innovative designs for O-DNP at high field strengths, using a gyrotron as a microwave source (see Sect. 3.2.3) and showed that reasonably high enhancements are feasible. Several other authors have polarized samples at lower fields in solution and transported these samples to higher fields, either by flow injection or by shuttling samples or the probe (see Sects. 3.2.1 and 3.2.2).

These concepts will be presented in Sect. 3.2. A third revolutionary concept for DNP arises from work by Ardenkjær-Larsen, who designed an apparatus to polarize samples at temperatures below 1.3 K, which are, after rapid dissolution, transported into a higher field magnet (D-DNP). This version of DNP achieves by far the largest enhancements by combining a Boltzmann and a DNP enhancement, although the irreversibility of the dissolution process limits applications to one brief NMR experiment to be carried out during the T_1 relaxation time of the involved nuclei. Despite this limitation, D-DNP has become a highly popular implementation of DNP, driven by enormous enhancements of >10,000 and considerable successes in MRI applications. D-DNP will be covered in Sect. 3.3. This review will not cover the design of DNP probes and microwave sources, which have been reviewed elsewhere [42].

In this review, polarizers are classified according to common DNP implementations, as introduced above, specifically O-DNP for liquids, SS-DNP and D-DNP. This is, of course, artificial because the same microwave and radiofrequency sources can be used in liquids or in solution, except for the use of different cooling arrangements, MAS or a dissolution device. This classification is, however, useful to describe current experimental implementations.

3.1 SS-DNP

As SS-NMR is covered by several other reviews by Griffin and coworkers [42, 180], the current state will only be summarized briefly. Early implementations of DNP for SS-NMR were described by Wind [83] and later by Schaefer [84]. Wind built a SS-DNP MAS spectrometer using a 1.4 T magnet (60 MHz proton frequency), 15 MHz ^{13}C frequency and an ESR frequency of 40 GHz [83] using a 10 W Klystron working at a fixed microwave frequency. He reported spectra of BDPA-doped polystyrene, where he achieved enhancements of 26 for the direct polarization of ^{13}C and 130 for a cross-polarization from protons to ^{13}C. He found that the polarization develops via a SE for ^{1}H and TM for ^{13}C [36]. Wind also examined a range of coal samples and diamonds. An alternative design was presented by Schaefer and coworkers, who observed an enhancement factor of 20 for polystyrene doped with the BPDA radical [84]. Griffin conducted similar experiments in 1993 using a gyrotron operating at 140 GHz and a newly designed DNP-MAS probe [85, 86]. He also found that he could obtain higher polarizations by cross-polarization from ^{1}H to ^{13}C, and concluded that TM yields higher polarizations than direct polarization via the SE [86].

The introduction of gyrotron masers by Griffin was clearly a landmark for DNP. Using nitroxide radicals (4-amino-TEMPO, see Sect. 3.4) he showed in 1997 the first MAS SS-NMR spectra of a protein, ^{15}N-Ala-labelled T4 lysozyme, with an enhancement of ~50 [38]. Griffin realized not only that polarizations via the CE are more efficient than those via TM, he also found that CE benefits from the close proximity of electrons in biradicals, leading to about four-fold larger signal intensities at much lower radical concentrations. He developed TOTAPOL, a water soluble biradical with a short linker [87, 88], which fulfills the need for a sufficiently strong dipolar coupling between the two electrons so that the difference between the electron Larmor frequencies approximates the nuclear Larmor frequency ($\omega_n = \omega_{e1} - \omega_{e2}$). With this concept, enhancements at 5 T and 90 K were increased from $\epsilon \sim 175$ to $\epsilon \sim 290$.

The search for optimal biradicals continues: more rigid radicals were shown to yield even larger enhancements [89], and recently a range of dinitroxides has been characterized [90]. Others have shown advantages for triradicals (DOTOPA-TEMPO) [91]. Thurber also demonstrated the advantage of lower temperatures for SS-DNP. The quest for even higher field strengths continues with gyrotron-based SS-DNP NMR spectrometers of up to 263 GHz (400 MHz proton frequency).

SS-DNP has become an important add-on to SS-NMR, with applications in protein NMR, as demonstrated for membrane proteins [92], amyloid fibrils [93] and an SH3 signalling protein [94]. However, the latter study by Oschkinat and Griffin also highlights the problems still encountered with SS-DNP for proteins. Deuteration was required to achieve any decent enhancement and the enhancement factor drops from ~70 to ~10 between 98 and 178 K. However, DNP at 98 K is impractical as the resolution at this temperature is insufficient for assignments.

Griffin has generally polarized protons with subsequent polarization transfer to ^{13}C. For many molecules this proved to be the right choice, despite the lower theoretical γ_e/γ_n ratio for protons (660) versus ^{13}C (2,625). However, Griffin's recent work shows that larger enhancements are possible by polarizing ^{13}C directly, although at the expense of a longer polarization time (~5 s for ^1H, 114–176 s for glycerol) [95], as a consequence of slower propagation of polarization for ^{13}C owing to the smaller dipolar coupling for ^{13}C compared to ^1H. Another option demonstrated by the same group is cross-polarization from ^2H to ^{13}C, which benefits from a large maximum enhancement ($\gamma_e/\gamma_{2H} = 4{,}300$) [82]. For proline, a polarization >700 was reported for ^2H using the Ox063 radical (see Sect. 3.4). This may become an interesting option for deuterated proteins.

Emsley and coworkers also reported surface enhanced NMR spectra for surface functionalized silica materials, where polarization is transferred from the protons of the solvent to the ^{13}C nuclei of functional groups at the surface, yielding at least a 50-fold signal enhancement [96].

As SS-DNP devices become commercially available we will most likely see widespread use across SS-NMR laboratories, and we may see results that render SS-DNP applicable for typical biological macromolecules.

3.1.1 Other Implementations of SS-DNP

There are now several implementations of gyrotrons, including a 260 GHz gyrotron oscillator (GYCOM, Russia) [94] and a 263 GHz gyrotron from Bruker (Germany), with higher frequencies announced. Idehara et al. reported the design of a Terahertz (980 GHz) gyrotron suitable for high-field NMR experiments at 20 T and beyond [97].

Recent reports covered the implementation of several solid state DNP devices without MAS, including one by Feintuch et al. operating at 95 GHz [98], operable over a wide range of temperatures. A high-power pulsed DNP spectrometer, which was used to demonstrate higher DNP enhancements using pulsed compared to continuous wave electron excitation, was recently described by Hunter et al. [99]. Such pulsed EPR solutions are likely to open new avenues. Thurber et al. presented a low-power 264 GHz solution operating at 7–80 K [100], and a solution for submicron solid state NMR imaging [91].

3.2 *O-DNP*

O-DNP has only recently seen a renaissance after the early experiments by Hausser [37] and later by Müller-Warmuth [39]. The design of such spectrometers for true liquid-state DNP is dictated by the need to minimize heating, the unfortunate field dependency and the desire to obtain high-resolution spectra. There are two possible options: (1) to polarize at low field and transport the sample to higher (or in selected cases to lower) fields, an implementation that has found many applications; and (2) to conduct the entire experiment at one field strength. The second option is currently targeted towards mechanistic studies, although such implementations may have considerable potential for future applications, if some fundamental problems can be solved. High field implementations suffer from the need to build a probe that yields high quality factors for both NMR and EPR and from the low predicted enhancements, although recent results are increasingly encouraging.

3.2.1 Low to High Field Transfer O-DNP

In the 1980s, Dorn explored DNP enhancements for flowing liquids, first with an arrangement of nearby microwave and radiofrequency cells, operating at 9 GHz and 14 MHz, respectively, in a 0.35 T magnet [101]. Subsequently, he designed an apparatus that flows the sample from a 0.34 T microwave cavity into a 4.7 T NMR magnet (Fig. 8) [101]. The decisive advantage of this arrangement is the separation of the microwave excitation from the radiofrequency excitation/detection probe, thus allowing for an optimal arrangement at either site. Dorn used TEMPO radicals either in solution or immobilized on silica gel [101, 103]. The latter case he termed SLIT (solid–liquid intermolecular transfer) for which he observed impressive enhancements for small molecules in organic solvents. The key advantages of SLIT are the avoidance of contact shifts and line broadening owing to the presence of large concentrations of the radical, facile detection of scalar-dominated signals, good transfer efficiencies owing to slow relaxation, and suitability for a wide range of flow rates [103].

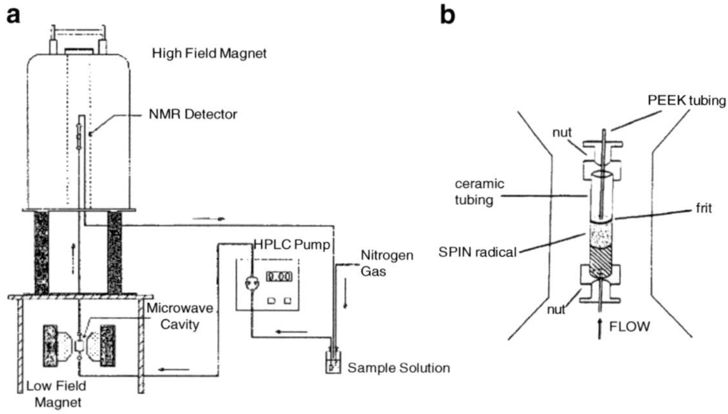

Fig. 8 (**a**) Dorn's O-DNP apparatus for continuous-flow online HPLC-DNP-NMR. (**b**) Diagram of the EPR flow cell employing TEMPO radicals immobilized on silica beads. [(**a**) kindly provided by Dorn; (**b**) reprinted from [102] with permission. Copyright 1994 American Chemical Society

For the case of dissolved radicals, Dorn found that much larger flow rates are required to obtain the optional enhancement, as a consequence of relaxation losses in the transfer [103]. Dorn derived a theoretical model to calculate the ultimate enhancements at low field [103], considering relaxation losses during the transfer, and achieved enhancements of -150 and -220 for benzene/dissolved TEMPO ^{1}H and ^{13}C, respectively. For chloroform/TEMPO he calculated extrapolated ^{1}H enhancements close to the dipolar limit of -330 and close to the scalar limit of 2,660 for ^{13}C for chlorinated carbons [104]. During flow these enhancements scale according to:

$$A_{\mathrm{obs}} = A \cdot B_{\mathrm{EPR}}/B_{\mathrm{NMR}} \cdot (1 - E_{1a})E_{1b}E_{1c}, \tag{21}$$

where $A = \xi f s \frac{\gamma_e}{\gamma_n}$ (the "Overhauser enhancement"), B_{EPR} and B_{NMR} are the EPR and NMR field strengths, and $E_{1a,b,c} = \exp(-t_0/T_{1a,b,c})$ is the relaxation in regions a, b and c of the apparatus shown in Fig. 8a. For Dorn's spectrometer, signals were scaled by a field factor of 14.4. With the immobilized free radical, he obtained under flow conditions enhancement factors of -56 for chloroform and 1–10 for benzene (dipolar mechanism). For ^{13}C he observed a reduced enhancement owing to three-spin effects involving protons for benzene, and factors of 40–60 for chlorinated carbons that polarize via a scalar mechanism [103]. Dorn reported a strong dependence of the polarization mechanism on the radical concentration for dipolar-dominated enhancements, with compensating scalar and dipolar enhancements at lower radical concentrations.

Dorn also proposed to use DNP-enhanced NMR as a detector for continuous-flow online chromatography [102] and later described a recycled-flow spectrometer [103]. The drawback of this is, however, the large dead volume of the pump and tubing that considerably reduces the overall enhancement when compared to a smaller volume sample recorded in a conventional probe [103], as shown in simulations of continuous flow NMR by Sudmeier et al. [105]. It would be worthwhile to reconsider the flow concept if the dead volumes could be further minimized.

The immobilized radical flow concept has recently been revisited by Lingwood et al. [106] to generate hyperpolarized water in the 0.35 T fringe field of a clinical 1.5 T whole-body MRI magnet, using an improved immobilized radical bound to Sepharose gel beads via polyelectrolyte linker arms. The system provides O-DNP enhancements approaching those of freely dissolved nitroxide radicals. This work reports a DNP signal enhancement factor of −15 in proton MRI images with respect to the unenhanced ^{1}H MRI signal of water at 1.5 T, reflecting an overall enhancement of −100 in the microwave cavity.

3.2.2 Shuttling DNP Spectrometer

A straightforward alternative to the continuous-flow concept is a shuttling solution, where either the sample or the probe is moved between magnets at two different field strengths. The concept of shuttling is quite old and was pioneered by Purcell and Pound in 1951 [107], who used it to examine the fate of a spin system upon field reversal. Several groups have implemented shuttling concepts since ([110–112], see [111] for a review on shuttling and field cycling).

Griesinger, Bennati and coworkers recently reported results from a concept shuttle spectrometer, polarizing at 9.7 GHz (0.23 T), on top of a 14.09 T magnet (600 MHz proton frequency). The maximum enhancement $I_{z,\max}$ depends on the actual DNP enhancement (8) at low field ($\epsilon_{\max}^{\mathrm{EPR}}$) and a field factor [56, 112]:

$$I_{z,\max} = \epsilon_{\max}^{\mathrm{EPR}} I_0^{\mathrm{EPR}} = \epsilon_{\max}^{\mathrm{EPR}} \left(B^{\mathrm{EPR}}/B^{\mathrm{NMR}}\right) I_0^{\mathrm{NMR}} = \epsilon_{\max}^{\mathrm{NMR}} I_0^{\mathrm{NMR}}. \tag{22}$$

The field ratio for the described implementation is ~41.4, i.e. a theoretical enhancement of −330 for a pure dipolar mechanism results, in the best case (for $\xi = f = s = 1$), in an overall NMR enhancement of ~8.

The apparatus shown in Fig. 9 [113] shuttles 0.9 or 0.46 mm inner diameter capillary samples within 140 ms between the magnet centres, which are 1.5 m apart. With a prototype version of this shuttle spectrometer an overall enhancement of −2.6 was observed for water doped with 5 mM TEMPONE-^{2}H,^{15}N [112], arising from an enhancement of −110 at low field. Much larger enhancements were observed for ^{13}C, +15 for $CHCl_3$ [112], in good agreement with large positive enhancements observed by Dorn. For 4 M urea, a negative enhancement with an $\varepsilon_{\mathrm{eff}}$ of -4 ± 1 was reported at radical concentrations of 25 mM [112]. Deuteration showed that three-spin effects (involving attached protons) play a minor role for urea. The low enhancement was therefore attributed to counteracting dipolar and scalar mechanisms.

In an improved setup (see [113] for a detailed description of the spectrometer and probe), relaxation-induced losses of hyperpolarization during the sample transfer were minimized by modifying the field profile so that the magnetic field never

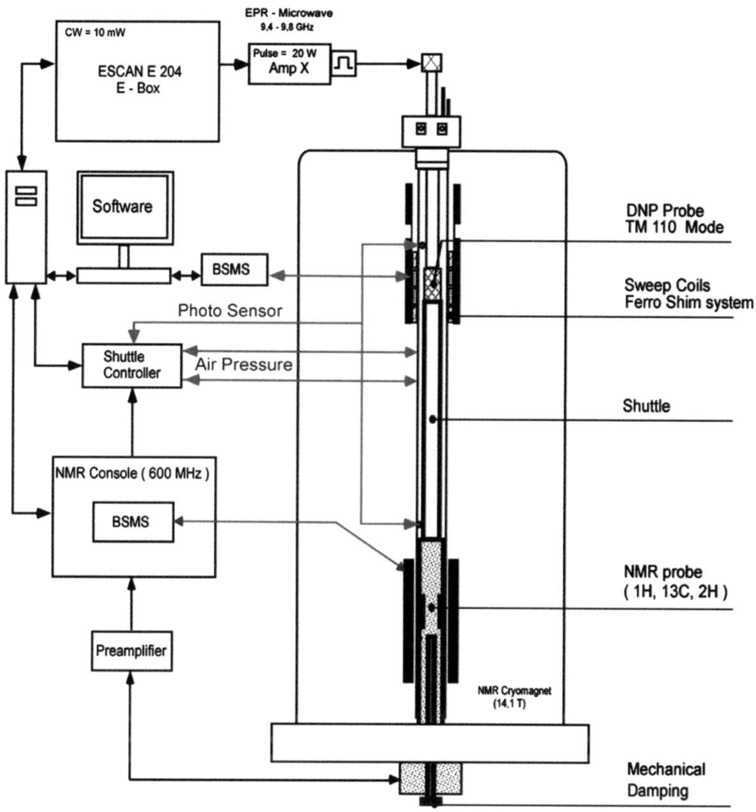

Fig. 9 O-DNP shuttling apparatus with an EPR spectrometer mounted on top of a 600 MHz magnet. An additional field plateau on top of the magnet is generated by a ferroshim system sitting in the stray field of the magnet. The EPR microwave cavity is connected to an EPR spectrometer. (Reproduced from [113] by permission of the PCCP Owner Societies)

dropped below the field strength of 0.34 T, at which hyperpolarization occurs (Fig. 10). An effective enhancement factor in the range of −1.4 to −2.8 was reported for the different protons of the D-glucose (10 mM $^{2}H,^{15}N$-TEMPONE solution in 99.8% D_2O with 5 mM DSS and 0.5 M D-glucose), reflecting relaxation losses for the glucose protons between 10 and 30%.

As pointed out by Griesinger and coworkers, losses during transfer associated with this arrangement are unfortunately still too large for protein applications [56]. Considering the field factor and relaxation losses during shuttling, any enhancement would be lost. To enable such a spectrometer for protein applications in solution therefore remains an unresolved challenge.

One major limitation of the shuttling implementations is, without doubt, the field factor between the EPR and NMR magnet. This could be reduced considerably with a 3.4 T (94 GHz) implementation of the polarizer, which would reduce the field

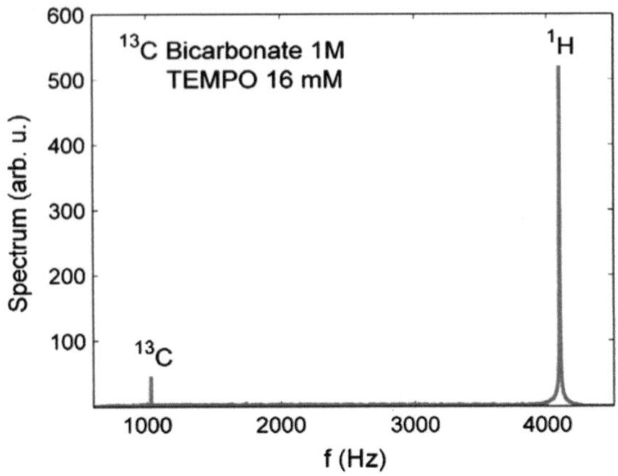

Fig. 10 Simultaneous measurement of proton and ^{13}C NMR signals of biscarbonate at 96 μT field using O-DNP [114]. Reprinted from [114], Copyright (2010), with permission from Elsevier

factor from ~41 to ~4.1. Such an arrangement would allow for larger enhancements, probably in the range of absolute values of ~10 for small molecules, considering smaller enhancements ε at larger field strengths. This is probably still not sufficient for protein applications, which would require much faster shuttling.

3.2.3 Stand-Alone O-DNP Polarizers

Many groups have implemented DNP spectrometers operating at microwave frequencies of ~9 and ~95 GHz, as the required hardware is readily and cheaply available. Many such systems are strictly add-on polarizers, some being used for MRI.

Münnemann described a mobile stand-alone polarizer using a Halbach permanent magnet operating at 0.30 T, corresponding to an electron frequency of ~9 GHz [115], designed to be used as a mobile DNP polarizer for clinical applications, but also to study enhancements at 3.2 MHz proton frequency in a relatively inhomogenous field. These authors compared different radicals: a trityl radical (TAM), TEMPOL (see Sect. 3.4) and a new spin-labelled cationic polyelectrolyte (poly(DADMAC)) with 4% nitroxide radical-bearing monomers and an average molecular weight of 490 kD. At high microwave powers, the DNP enhancement for water using poly(DADMAC) significantly exceeded those of TEMPOL. Enhancement values of −80 for poly (DADMAC), −60 for TEMPOL and −27 for the trityl radical were reported.

Han and coworkers presented several implementations of portable polarizers operating at 0.25 T/ 9.5 GHz microwave frequency [116]. In a first stand-alone solution using a fixed field electromagnet and alternatively a field-adjustable Halbach magnet, the authors reported an enhancement factor of −110 using the electromagnet for water and TEMPOL as a polarizer. For the more portable Halbach magnet, they achieved only a factor of −65, a consequence of a less optimal space-constrained microwave cavity.

3.2.4 DNP in the Earth's Field

It should be mentioned that μT earth field MRI is developing its own momentum. On the detection side, these experiments benefit from superconducting quantum interference devices (SQUID) [117], owing to the frequency independence of the SQUID devices. Earth field measurements were often combined with prepolarization in magnets with modest field strengths, benefiting from huge enhancements if combined with DNP, in particular if the polarization is carried out at higher field strengths (see [118] for a thorough theoretical treatment). Very low-field implementations represent an interesting theoretical challenge, as the Hamiltonian is not at all diagonal in product operator representations. For such low fields the hyperfine coupling term cannot be treated as a perturbation to the dominant Zeeman terms, but becomes strongly dominant.

A recent implementation by Halse, measuring in the Earth's field strength of 54 μT using nitroxide free radical microwave irradiation at 131.5 MHz in the presence of a 2.7 mT prepolarizing field was shown to yield an enhancement factor of −2,250 over thermal equilibrium [119]. Halse et al. also showed a ^{19}F–^{1}H COSY acquired of neat 2,2,2-trifluoroethanol in the Earth's magnetic field.

A more recent apparatus designed by Zotev et al. ([114], also see references included in this article) uses SQUID detectors and 96 μT, corresponding to the Larmor frequency of 4 kHz for protons and 1 kHz for ^{13}C. O-DNP was carried out with nitroxide radicals at 3.5–5.7 mT fields using radiofrequency irradiation at 120 MHz. The authors demonstrate applicability for imaging, but also for spectroscopy of ^{13}C-labelled metabolites such as sodium bicarbonate (Fig. 10), pyruvate, alanine and lactate. Enhancement factors or −95 for protons and −200 for ^{13}C, corresponding to thermal polarizations at 0.3 and 1.1 T field strengths, respectively, were reported. Such implementations could have significant impact for metabolic imaging, as an alternative to D-DNP, especially as NMR lines become very narrow at low fields, even for significantly inhomogeneous fields [117].

3.2.5 O-DNP Operating at a Single Field Strength for EPR and NMR

The challenge of O-DNP is undoubtedly to enable DNP in liquids at high magnetic fields, ideally at 600 MHz, or at least at 400 MHz proton frequencies. Such an implementation would eliminate the field factor, which lowers the enhancement for shuttle spectrometers. This requires a design of co-localized microwave and radiofrequency setup, ideally enabling high-resolution NMR spectra. One question is how much the enhancement drops at higher fields, further away from the $\omega_e \tau_c \ll 1$ condition. The design of such a DNP-NMR spectrometer has now been tackled by several groups, with implementations of up to 260 GHz. Although the overall challenge of detecting biological macromolecules in solution has not yet been met, there is some hope that this may become feasible in the future.

Recent fundamental work to identify the factors determining the efficiency of the polarization process (s, f and ξ) was primarily conducted by the groups of Bennati, Prisner, Kengens, Dupree and Han. Some of this has already been discussed in Sect. 2. A recent issue of *Physical Chemistry Chemical Physics* includes several articles examining O-DNP enhancements at 9.7, 94 and 260 GHz. Several studies show larger enhancements than expected from molecular dynamics simulations by Sezer [58]. These experiments along with the experimental implementations will be summarized next.

Bennati and coworkers examined experimentally the factors that determine efficient polarization at high fields, and compared enhancements at 9.7 and at 94 GHz [120]. For this purpose, two different DNP setups were used, both based on commercially available (Bruker) EPR spectrometers and ENDOR probes. For the 94 GHz DNP spectrometer, a 400 mW power upgrade was installed. The groups reported an O-DNP enhancement ε of -170 for water at 9.7 GHz (15 MHz proton frequency), using TEMPONE-D-^{15}N, the highest observed so far (Fig. 11a). However, at 94 GHz (140 MHz proton frequency) only an enhancement ε of -43 was achieved. These results showed that previous measurements by the same group yielding enhancements of -140 [121] were power-limited. A systematic study looking at the field dependence of the saturation factor, and measuring the temperature of the sample using different methods for different sample sizes, showed that saturation factors are near unity and coupling factors ξ are ~0.36 for radical

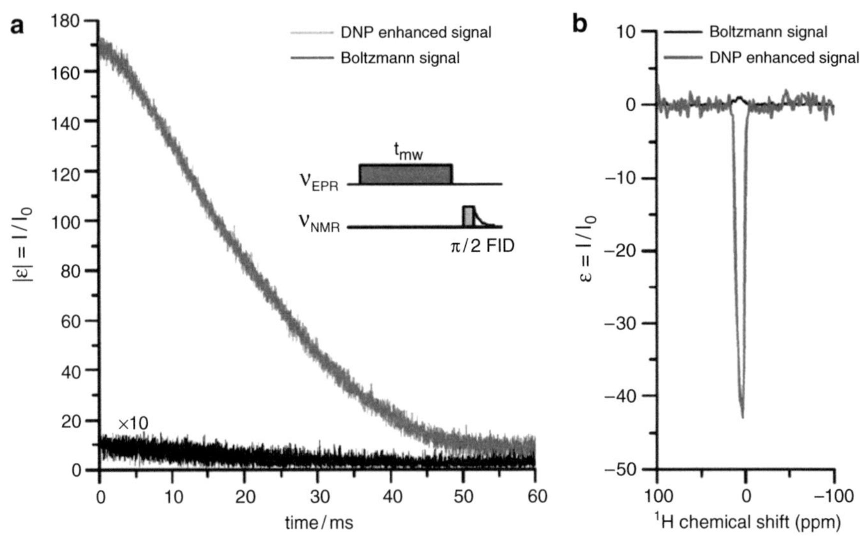

Fig. 11 O-DNP experiments by Bennati and coworkers. (**a**) Magnitude of the NMR free induction decay (FID) for 25 mM D-^{15}N-TEMPOL in water with and without microwave irradiation (9.7 GHz), showing a maximum enhancement ε of -170 for the first point of the FID. The Boltzmann signal required 4,096 scans whereas the DNP-enhanced signal was recorded within eight scans. (**b**) ^{1}H NMR spectra of water containing 25 mM D-^{15}N-TEMPONE at 94 GHz. (Reproduced from [45, 120] by permission of the PCCP Owner Societies)

concentrations of 5–10 mM and drop to 0.33 for 25–50 mM. These values were also found to be in good agreement with preceding NMRD measurements [54] and molecular dynamics (MD) calculations [58] but discard coupling factors determined by Armstrong et al. [55]. Saturation factors were ~0.8 when considering both lines and almost 1 when considering only one EPR line. At 94 GHz, the effective saturation factor s was estimated at 0.65, using a leakage f of 0.91, and a coupling factor ξ of 0.11 from MD measurements. It is therefore mainly the field dependence of ξ that is responsible for the lower enhancement at higher fields.

Kentgens and coworkers conducted a similar study using a DNP probe based on a novel double-resonance structure combining a single-mode microwave cavity resonating at 95 GHz with an intra-cavity radiofrequency coil operating at 95 GHz and 140 MHz [122]. These authors reported enhancements ε of -65 and extrapolated those to -94 in the centre of the resonator. These values are considerably higher than those reported by Bennati and coworkers; they are also higher than expected from MD calculations, even when some sample heating is considered.

A third implementation of high-field O-DNP was developed by Prisner and coworkers [123, 124]. In fact, they compared two high field implementations at 260 GHz and 400 MHz, using two different 260 GHz microwave sources, a high-power gyrotron microwave source (maximum power up to 20 W) and a solid-state microwave power source (maximum power 45 mW). Their setup uses a resonant microwave structure to separate electrical and magnetic field components for lossy solvents like water, and to increase the magnetic field strength at the sample (described in detail in [123]). The disadvantage of this setup is a very poor radiofrequency homogeneity, leading to broad NMR resonances, and the requirement for tiny capillary samples (0.03 mm inner diameter with few nanolitres of volume).

Prisner and coworkers started to use Fremy's salt as a radical, which exhibits a narrow EPR linewidth of less than 0.1 mT at 260 GHz [124] and yields much higher enhancements (-10 compared to -6) for ^{15}N-TEMPOL [122]. The gyrotron microwave source exhibited a striking improvement over that previously reported for the 45 mW source, yielding an enhancement value of -29 on water protons using Fremy's salt, compared to -10 previously reported for the 45 mW source (Fig. 12).

Denysenkov et al. discussed in detail the implications of this enhancement, which exceeds all expectations from preceding MD calculations [123]. The leakage factor was determined to be $f = 0.94$. The saturation factor s, which is more difficult to obtain owing to the short electron relaxation times T_{1e} and T_{2e}, was estimated computationally using Redfield's perturbation theory [125]. However, the variation of s within the possible window of T_{1e} of 100–400 ns cannot account for the high ε. For a coupling factor ξ of 0.072 estimated from MD calculations [58], a saturation factor of $s = 0.65$ was estimated, which is higher than expected from simulations. Such higher than theoretically expected enhancements can either be attributed to higher saturation factors or to incorrect temperature measurements, as suggested in [59].

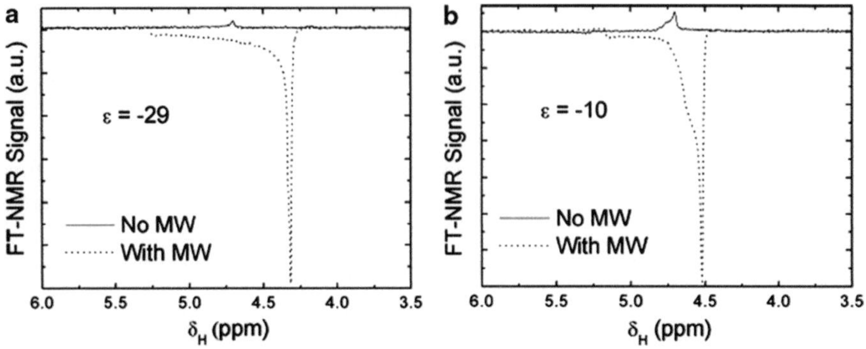

Fig. 12 Water ^{1}H NMR spectra reported by Prisner and coworkers, conducted at 9.2 T (260 Hz, 400 MHz); acquired with (*dots*) and without microwave pumping (*solid line*) on 40 mM aqueous Fremy's salt solutions using (**a**) a high power gyrotron, and (**b**) a low power solid-state microwave source. (Reproduced from [123] by permission of the PCCP Owner Societies)

Finally, Kryukov et al. designed an apparatus with a 100 W microwave amplifier operating at 94 GHz, using a modified Bruker ENDOR probe [126]. He examined TEMPOL in toluene, the only study looking at a non-aqueous solvent, and observed unexpectedly high enhancements of −43 for the methyl group and −50 for the ring protons [126]. A maximum enhancement was reported for a 20 mM concentration of TEMPO, similar to Bennati's results in water [120]. Again, enhancements were significantly greater than expected from extrapolation of early lower-field experiments, questioning preceding theoretical evaluations.

Direct observation of ^{13}C may be an option to overcome some of the limitations encountered for ^{1}H-DNP. This has already been demonstrated by Hausser at low field [37], and later by Loening et al. [127] at a magnetic field of 5 T (211 MHz for protons, 140 GHz for electrons), who reported an enhancement of −40 for tetrachloride (besides other heteronuclei). More recently, Lingwood et al. reported values of −265 for urea, +60 for *N,N*-dimethylformamide and −160 for pyruvate. Reese et al. reported effective high-field enhancements of +15 and −4 for chloroform and urea, respectively, using the earlier described shuttle spectrometer (see Fig. 13 for $CHCl_3$) [56]. The limitation for ^{13}C is the complex mechanism arising from competing scalar and dipolar effects, which may even compensate.

3.2.6 Applications of O-DNP

The overall picture arising from these implementations and applications suggest increased potential of O-DNP in solution. However, heating remains a serious problem that will limit applications, particularly for heat-sensitive biological macromolecules. The small sample sizes required in all implementations represent a serious practical obstacle. Large enhancements reported for higher fields suggest that a 94 GHz shuttle system may represent a worthwhile advantage over the existing 9.7 GHz implementation because the smaller field ratio will dominate over the

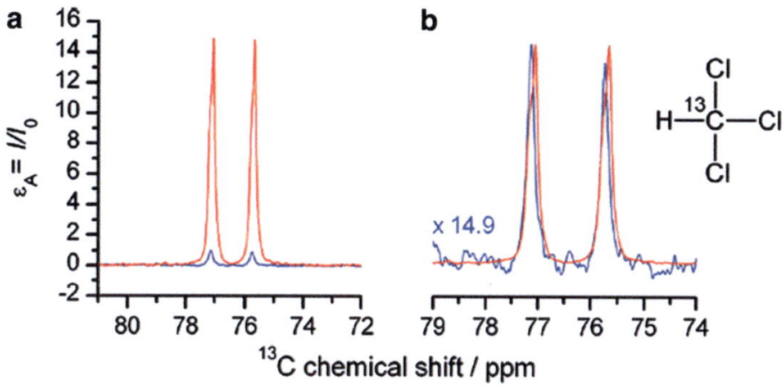

Fig. 13 ^{13}C (shuttle) DNP enhancement reported by Reese et al. [56] of ^{13}C-chloroform in water with 25 mM D-^{15}N-TEMPONE. The enhanced signals are compared with the Boltzmann polarization at 14 T. Reprinted with permission from [56]. Copyright 2009 American Chemical Society

somewhat reduced enhancement, predicting enhancements of at least 10 for protons. However, applications to biological macromolecules are currently limited by heating and shuttling times, which are still too long for protein relaxation times. So far, all measurements conducted at the same field strength for EPR and NMR suffer a from poor B_0 homogeneity, whereas shuttle spectrometers benefit from using a standard NMR probe to record high-resolution NMR spectra. Dorn's experiments pointed towards useful analytical applications of O-DNP, although hampered by the large amount of solvent required for the flow arrangement. If revisited, a much smaller dead volume and fast flow rate would have to be considered.

Several applications of O-DNP were suggested by Han and coworkers, who proposed hyperpolarized water as an MRI contrast agent, and applied this for the visualization of vortices in model reactors [51]. For in vivo MRI applications this approach is, however, limited by the relatively small enhancements for water, the requirement of using relatively toxic TEMPO radicals, and the fast relaxation of water (~1.5 s), leaving little time between the polarization, the administration of the polarized water and the MRI measurement. The same group also used DNP to study the dynamics of water at lipid vesicle surfaces [128], and more recently to study site-specific hydration translational dynamics in the core of a protein molten globule, using site-specific nitroxide spin labels [129]. The study shows that the translational dynamics of bound water in the nonpolar core of the molten globule of apomyoglobin is only four- to sixfold slower than that of bulk water. At this point such indirect measurements seem to be the only avenue towards the use of O-DNP in studies on biological macromolecules.

3.3 Dissolution and Temperature Jump DNP

Dissolution DNP and temperature jump DNP are two related concepts that share the common principle of carrying out the DNP enhancement at low temperatures, followed by sample heating to prepare the sample in the liquid state for the acquisition of an NMR spectrum. Dissolution is achieved employing either hot solvent or laser heating. These implementations combine the power of carrying out NMR experiments in solution with the much larger DNP enhancements achievable at low temperatures.

3.3.1 Dissolution DNP

Dissolution DNP (also ex situ DNP), pioneered by Ardenkjær-Larsen [44], yields by far the largest overall enhancements of all implementations of DNP. This is achieved by polarizing in a lower field magnet (typically 3.4 T, microwave frequency 94 GHz) at 1.1–1.5 K, followed by rapidly melting and dissolving the hyperpolarized sample using hot solvent. The sample is then rapidly transferred into a conventional NMR magnet, to record the NMR spectrum ex situ at room temperature. This implementation combines a massive Boltzmann temperature enhancement of 200–250 with the actual DNP enhancement ε. The overall enhancement is defined by:

$$\epsilon' = \epsilon \cdot \frac{B_{\mathrm{DNP}}}{B_{\mathrm{NMR}}} \cdot \frac{T_{\mathrm{NMR}}}{T_{\mathrm{DNP}}}. \tag{23}$$

For a 11.75 T NMR system (500 MHz proton frequency) the B ratio amounts to 3.45. This is small compared to the huge temperature factor of ≈ 250. Even for modest DNP enhancements of $\approx$150, overall enhancements of >10,000 are obtained.

Figure 14 shows a scheme of the overall arrangement. This setup has found widespread use through the availability of a commercial implementation by Oxford

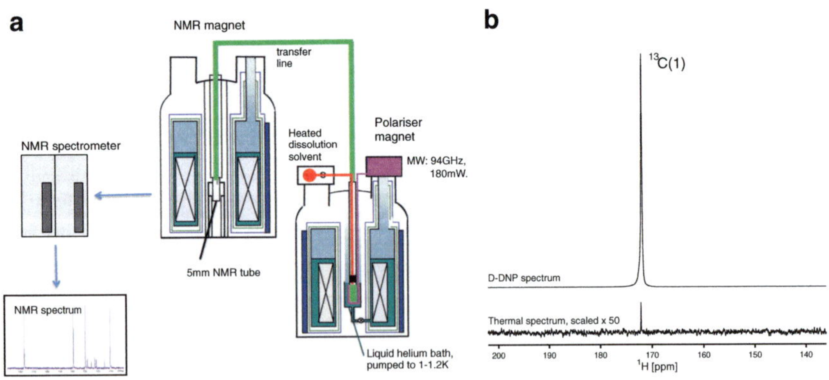

Fig. 14 (**a**) D-DNP setup. The sample is polarized at 1–1.3 K in the polarizer (shown on the *right*) and is subsequently transferred to the NMR magnet (*left*), where an NMR spectrum is recorded. Further details of this design can be found in [44]. (**b**) ^{13}C-NMR spectrum of ^{13}C(1)-pyruvate after D-DNP versus unpolarized, the latter vertically scaled by a factor of 50, showing the immense enhancement obtained by D-DNP

Instruments termed Hypersense^{TM} [130], and due to the development towards clinical use by GE Healthcare [131]. There are, however, several alternative implementations, including a similar concept using a smaller He compartment [132], and a sophisticated solution using a two-centre magnet, designed by Köckenberger and coworkers [133].

The inevitable drawback of this implementation of DNP is the irreversible melting and dissolution of the sample, which limits the subsequent acquisition of NMR spectra to the life-time of the nuclear spins, i.e., the individual longitudinal relaxation times. While this seems to represent a major drawback, the huge enhancements observed for D-DNP may well compensate the loss of signal averaging, especially considering the availability of fast multidimensional acquisition schemes.

Experimental Conditions of D-DNP

As for SS-DNP, samples for D-DNP are prepared in a glassy matrix; crystalline (at least microcrystalline) samples do not yield any comparable polarization. Glasses can be readily prepared using mixtures of glycerol, DMSO, methanol and water. The Hypersense polarizer uses sample volumes of maximum 250 μL. A temperature of 1.1–1.3 K is reached by pumping on the He bath, regulating the He supply from the magnet's main He tank by a needle valve. The dissolution process is achieved using hot pressurized solvents, typically at 8–10 bar and ~200 °C, specific conditions depending on the dissolution solvent. After dissolution, samples are transferred into the NMR magnet through a tube by using gas pressure. Different dissolution solvents have been used, but mostly water and methanol, where the addition of methanol improves the sample flow and reduces bubble formation in the transfer process. To reach room temperature by dissolution it is usually necessary to use ~4 mL of solvent, of which only a fraction is transferred into the NMR tube (particularly when 5 mm NMR tubes holding 0.5 mL are used). Careful calibration of the solvent pressure and temperature guarantees a reasonably reproducible sample temperature after dissolution (± 1 °C). The dissolution and transfer process each take 2–3 s, and this time frame restricts application to samples with sufficiently long relaxation times in the dissolution buffer. Hilty developed a sample injection device that enables significantly faster and more reliable sample injection in a time frame of ~600 ms [134]. With this arrangement, Hilty was able to record high quality spectra of glucose. Considering that spectra are recorded in a short time frame, pre-tuning and pre-shimming on standard samples yields reasonably good NMR spectra in unlocked mode.

The presence of the radical in the final sample (0.2–0.3 mM) affects relaxation times. Several methods have been described to remove the radical, e.g., filtration through a short anion-exchange column [44], radical quenching using a reducing agent [135], and precipitation under acidic conditions followed by mechanical filtration [131]. Removal of the radical is particularly important for in vivo applications because radicals are likely to be toxic, but also for analytical applications of D-NMR, to minimize relaxation losses during the transfer from one magnet to the other.

D-DNP has had great success through the in vivo application of metabolic imaging, which holds great promise for clinical diagnostic applications. One of these in vivo applications is the detection of the fate of ^{13}C(1)-pyruvate in tumours, which characterizes the metabolic state of a tumour by the ratio of the metabolization products alanine and lactate [136, 137]. Another important in vivo application is the measurement of pH in vivo using hyperpolarized ^{13}C-labelled bicarbonate [138]; this can be used to image pathological processes associated with alterations in tissue pH, such as cancer, ischaemia and inflammation. The number of applications of metabolic imaging is growing quickly; the field deserves a separate review and will therefore not be covered here (see for example [139, 140]).

Polarization Mechanisms in D-NMR

Ardenkjær-Larsen first examined the polarization of ^{13}C-urea in glycerol with OX063 as a trityl radical (see Sect. 3.4) at 1.1 K. He observed a build-up constant τ of 4,900 s, while the T_{1n} of the ^{13}C signal was reported to be 28,200 s. For urea, a ^{13}C polarization of 42% was reached using15 mM OX063 at a temperature of 1.1 K. At 20 mM OX063, the polarization was reduced to 26%, and τ and T_{1n} were shortened to 2,755 and 15,800 s, respectively [44]. Based on microwave sweeps (polarization vs. microwave frequency curves) the ^{13}C polarization mechanism was attributed to thermal mixing, although a contribution by the SE could not be excluded. In the light of a rather lengthy polarization process, the SE seems to be more likely.

A more detailed examination of the polarization process of pyruvic acid at 1.1 K with 15 mM OX063Me [141] showed that T_{1e} is rather long at 0.91 s, and $T_{1n} \approx$ 12,000 s [for the pyruvic acid ^{13}C(1)]. T_{1n} also depends on the concentration of the radical ($T_{1n}^{-1} = 1.1 \cdot [\text{mmol/l}]^{-1} \cdot c + 82 \times 10^{-6}\text{s}^{-1}$). In the same work from the Ardenkjær-Larsen group, the effect of the addition of Gd^{3+} was examined. $GdC13_3$ dramatically shortened T_{1e} at concentrations of 1 mM but not T_{1n}, and increased the achievable ^{13}C polarization, but not the build-up time.

Many aspects of the underlying mechanism remain enigmatic, and recent measurements show that the current level of understanding cannot even explain basic properties such as the field dependence of the polarization. For ^{13}C, a TM mechanism has been postulated considering that the homogenous width of the EPR signal is larger than the Larmor frequency ($\omega_{e1/2} = 63\text{MHz} > \omega_n = 35.88\text{MHz}$). Reynolds examined build-up curves and microwave sweeps for various nuclei, including ^{2}H, ^{31}P, ^{15}N, ^{13}C and ^{29}Si to determine the polarization mechanism in each case [142]. He measured the frequency separation of the opposite sign maxima and compared those to the Larmor frequencies of the nuclei (according to $\omega = \omega_e \pm \omega_n$ the separation should be $2\omega_n$ for the SE). All nuclei with ω_0 smaller than that of ^{13}C (35.96 MHz) were found to polarize via TM. This includes ^{15}N, ^{2}H and ^{29}Si. Only ^{31}P seems to polarize via a poorly resolved SE because its Larmor frequency

(58.875 MHz) is larger than the electron line half-width $\omega_{e1/2}$. Reynolds also observed a smaller secondary signal attributed to a double SE at a separation of about 147 MHz, attributed to a joint electron-heteronuclear-proton spin flip. Considering these frequency dependencies, it should be possible to polarize multiple nuclei with the same microwave frequency, which has in fact been demonstrated for ^{13}C and ^{15}N by Day et al. [143].

These findings raise the question whether the choice of a low temperature and a trityl radical enforces a TM mechanism. If this was the case it should be possible to identify conditions similar to those for CE in SS-DNP under which the effect could be optimized by using radicals with a suitable coupling between manifold of spins. Ardenkjær-Larsen and coworkers therefore explored trityl biradicals [144]. However, for a panel of biradicals the overall enhancement was lower than for monoradicals and was associated with a longer build-up time, although at lower optimal radical concentrations. No optimum trityl–trityl distance could be identified. Curiously, the ^{13}C T_{1n} was shortened for the biradicals, whereas T_{1e} was unaffected by radical type (and field strength).

However, both mono- and birdicals showed a doubling of the polarization when increasing the field from 3.35 to 4.64 T [141, 145]. Higher polarizations at higher field strength (5 T) were also reported by Jannin et al. [146]. This observation is clearly in disagreement with the expected B_0^{-1} dependence. Strangely, the enhanced polarization Gd^{3+} was not observed at the higher field of 4.64 T [141, 145]. These results make little sense in the light of existing theoretical models of either the SE or TM, and highlight a limited understanding of the polarization mechanism, even for simple systems.

Moreover, most biological applications of D-DNP use trityl radicals in pure pyruvic acid, void of any solvent addition. Compounds in smaller concentrations polarize at a much lower level, and this polarization level varies vastly between different substances. Some substances show little or no polarization despite a long T_1 (for example citrate, despite the long T_1 of its quaternary carbon) [147]. Polarizability strongly depends on the contact between the radical and the molecule to be polarized [147], and this contact is modulated by the solvent. Polarizability also depends fundamentally on the polarization matrix, in particular on efficient spin diffusion [43]. This is highlighted by studies showing optimal polarization for many small molecules in deuterated solvents, and in the presence of substances that improve spin diffusion by enhancing the network of ^{13}C atoms. As shown by Ludwig et al., higher enhancements can be achieved for many small molecules polarized in dilute solutions if ^{13}C(2)-acetone is added to the polarization matrix as a co-polarization agent [66]. This effect was confirmed by Lumata et al., who observed faster polarization of pyruvate by adding ^{13}C-DMSO to the polarization matrix [67]. Similarly, solvent deuteration tends to reduce polarization times by reducing relaxation mechanisms that influence spin diffusion in the glassy sample matrix.

Moreover, for substances bearing methyl groups interference with quantum tunneling has been observed [70]. These arise from the sudden cooling, causing a skewed equilibrium for tunnel transitions in methyl quantum rotors.

Reestablishment of the equilibrium population is associated with changes in the spin state, through the influence of the exclusion principle on the space-spin symmetry of the molecular states. As a consequence, the signals of methyl ^{13}Cs may appear with different sign compared to the other resonances. As a consequence, the absolute intensity of the spectrum is not the same any more for microwave frequencies of $\omega_e - \omega_n$ versus $\omega_e + \omega_n$, because only the latter cause constructive interference of quantum rotor and DNP effects. It should be mentioned that quantum rotor mechanisms constitute a separate polarization mechanism with great potential [13, 15, 16].

Implementations of D-NMR

The immense enhancements ϵ' obtained for D-NMR have spurred several alternative implementations. A polarizer similar to the one described by Ardenkjær-Larsen has been designed and built by Comment et al. [132, 148]. He and his coworkers also explored polarization build-up at 1.2 K using TEMPO rather than trityl radicals [149], and found that a tight packing of spherical beads of 2–3 mm diameter, obtained by dripping the solution into liquid nitrogen, leads to higher polarizations, probably reflecting better microwave penetration into the sample. The highest polarization, of 14% or >15,000, reported by these authors was for 3 M acetate in a 9:1 mixture of D_2O and ethanol doped with 33 mM TEMPO. The main DNP mechanism was thought to be TM.

Köckenberger and coworkers designed a D-DNP polarizer using a dual-isocentre superconductive magnet with isocentres at 3.35 T and 9.4 T (400 MHz) separated by 85 cm [133], based on a magnet designed by Oxford Instruments. In this design, the polarizer system is placed on top, with the waveguides fed in through the top of the magnet. This system was designed to allow for sample transfer in the solid state, dissolving the sample only in very close proximity to the site where the NMR experiment is performed. The minimum magnetic field strength between the two isocentres is 0.2 T, thus avoid passage through very small or zero field. The system exhibits a transfer time and settling time of 2.9 s, about half of that achieved with conventional D-DNP systems. This system was characterized using an Ala–Gln dipeptide (Fig. 15) and a pentapeptide. Köckenberger and coworkers reported impressive results for these samples. Figure 15 shows the ^{1}H and ^{13}C spectra of the dipeptide, with enhancement factors for each ^{1}H resonance line for the dual centre magnet versus a stand-alone polarizer. An ϵ' of almost 3,000 was observed for C7 of Ala–Gln despite its rather short T_1 time constant of 700 ms.

Two-Dimensional Spectra Using D-DNP

The limited life-time of the polarization and the irreversibility of the dissolution process impose strict limitations on the types NMR of experiments that are feasible. Commonly, NMR experiments are recorded over longer time periods by averaging

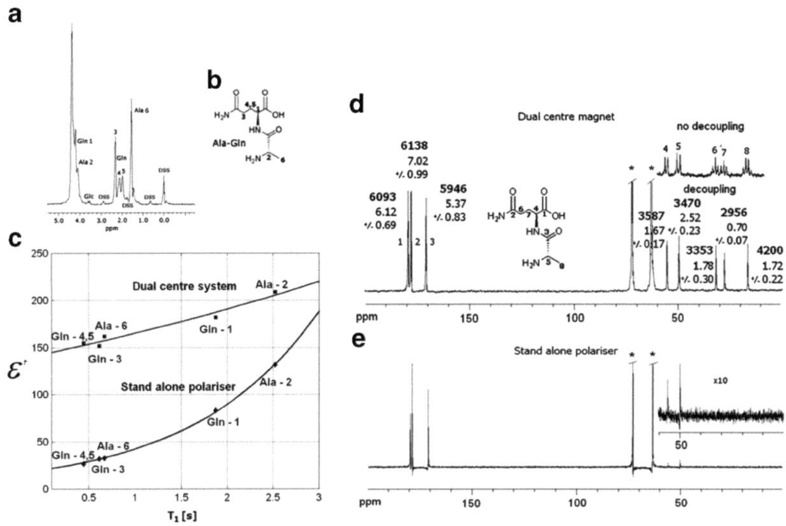

Fig. 15 (**a**) ^{1}H spectrum of the dipeptide Ala–Gln (**b**) after polarization at 3.4 T, shuttling in the solid state between isocentres of a dual-isocentre magnet and consecutive dissolution. (**c**) Comparison between enhancement factor ϵ' for the ^{1}H resonances of Ala–Gln after polarization in the dual-isocentre instrument versus using a stand-alone polarizer with pneumatic transfer. (**d**) Comparison of single scan ^{13}C spectra of a 0.5 mM solution of the Ala–Gln dipeptide (**b**) for the dual-isocentre DNP-NMR spectrometer versus a stand-alone polarizer. *Bold numbers* give the enhancement factor ϵ'. In addition, the T_1 relaxation time constants of the corresponding resonance lines are shown. (Reproduced from [133] by permission of the PCCP Owner Societies)

of larger numbers of scans. This is not possible with D-DNP because the NMR experiment has to be recorded within the life-time of the hyperpolarization.

To overcome this limitation, several authors have developed schemes to record two-dimensional spectra within the time frame imposed by the T_1 relaxation time constants of the molecules. Such experiments include gradient-encoded single-scan two-dimensional spectroscopy (ultrafast two-dimensional NMR) by Frydman and coworkers [150, 151] (Fig. 16a). Köckenberger used the ultrafast concept to implement a slice-selective single-scan proton COSY, demonstrated for a small peptide as shown in Fig. 16b [152]. Hilty used off-resonance decoupling to reconstruct a second dimension from a series of one-dimensional experiments [154], and later small flip angle excitation HMQC spectra [155]. Small flip angle single-run HMQCs were also used by Ludwig et al. [66], who combined this concept with non-linear sampling to improve the resolution in such spectra (Fig. 16c, d).

Storage of Hyperpolarization

Considering the short T_1 relaxation time of some nuclei, it seems straightforward to explore longer-lived tags for D-DNP experiments. One of the options are acetyl

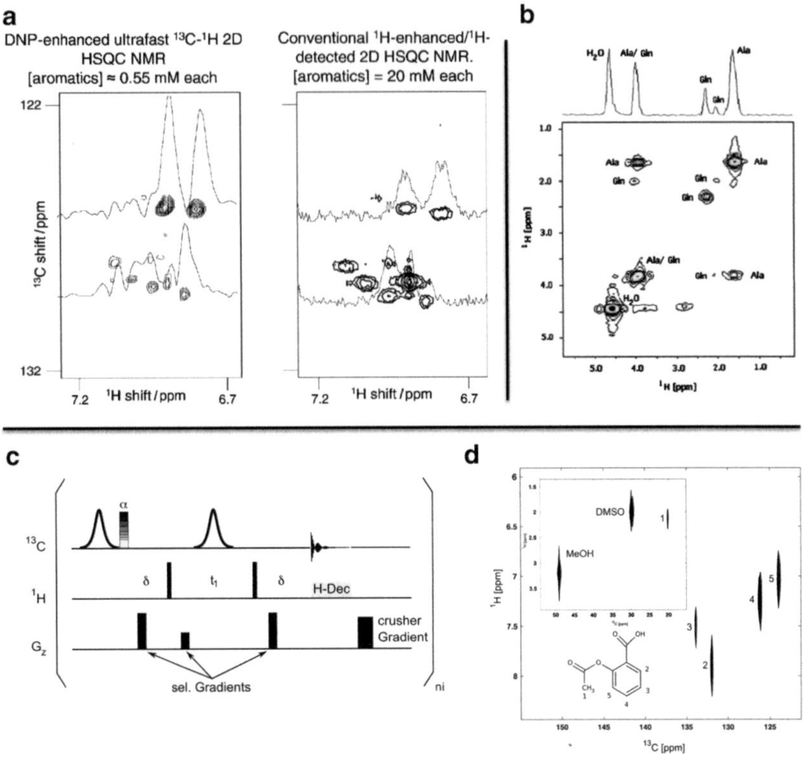

Fig. 16 Two-dimensional spectra from different authors, all using D-DNP and fast acquisition schemes. (**a**) ^{13}C-^{1}H-HSQC NMR spectra of a 1:1:1:1 mixture of *o*-, *m*-, and *p*-xylene and toluene, comparing a spectrum recorded by Frydman's ultrafast method (*left*) with a conventionally recorded spectrum (*right*) [150]. (Copyright Wiley-VCH. Reproduced with permission). (**b**) DNP enhanced proton COSY spectrum of 1.5 mM Ala–Gln at 9.4 T using an ultrafast COSY method by Köckenberger and coworkers. (Reproduced from [151] by permission of the PCCP Owner Societies). (**c**) 2D small flip angle ^{13}C-^{1}H-HMQC pulse sequence by Ludwig et al. [66, 153]. (**d**) 2D-HMQC spectrum of aspirin recorded in 30 s with 64 increments after 90 min of polarization at 1.3 K (kindly contributed by C. Ludwig)

tags, which can be attached to many small molecules near OH or NH_2 groups, and represent excellent substrates for D-DNP considering their relatively long T_1 relaxation time (>10 s at 11.7 T and 37 °C). The utility of acetyl tags for DNP was first realized by Wilson et al. [156] who used it to resolve a mixture of amino acids Gly, Ser, Val, Leu and Ala, for several small peptides and for *N*-acetylcysteine. Signal enhancements of up to 1,400-fold were reported for these molecules. It should be noted that this approach is limited to those amino acids that are sufficiently soluble before and after acetylation. Ludwig recently showed that ^{13}C-labelled acetyl tags produce a ^{13}C-^{13}C-NOE over a relatively long time that can be used to enhance ^{13}C signals after dissolution [153].

Bodenhausen and coworkers explored long-lived singlet spin states (LLS) to preserve hyperpolarization [157]. As shown by Levitt and coworkers, LLSs are delocalized on two or more coupled spins and have unusually long relaxation times because they are not affected by intramolecular dipole–dipole relaxation [158–160]. They can therefore be used to store magnetization, as demonstrated by Vasos et al. for the dipeptide Ala-Gly [157]. For this, the natural abundance dipeptide was polarized, the polarization on long-lived ^{13}CO carbons ($T_1 \approx 26$ s) in Gly was transferred via inverse INEPT to the Gly H^{α} protons, sustained over a period of time as long as $5T_{LLS}$, and finally transformed into detectable proton magnetization. The life-time of the LLS in Ala-Gly was reported to be $\approx$42 s.

Warren showed that in selected molecules such as 2,3-carbon-^{13}C–diacetyl with chemically equivalent nuclei, spin states with relatively long lifetimes exist that may be converted into detectable magnetization through a hydration reaction [161].

In principle, one of these options could be used to design tags, using either long-lived singlet spin states or simply groups with long T_1s, such as acetyl tags exhibiting T_1s of 20–30 s. For LLS, it would of course be preferable if the LLS could be generated in the polarizer to store polarization during the transfer process.

Applications of D-NMR

Despite the quasi "one-scan limitation" of D-NMR, the method has produced a multitude of interesting applications, including time-resolved real-time measurements of enzyme kinetics [162, 163], the analysis of biosynthetic pathways [164], the detection of low-populated reaction intermediates with hyperpolarized NMR [165], and ligand detected protein–ligand interactions [166]. Besides those biochemical applications, there is already a large base of literature on metabolic analyses and metabolic imaging (see for example [137, 139]). The existing work clearly demonstrates the applicability of D-DNP for chemical, biochemical and medical applications. A major drawback lies in the requirement of massive cooling, requiring ~1 L of liquid He per hour, indicating a need for more effective polarizers, possibly using smaller sample volumes.

3.3.2 In Situ Temperature-Jump DNP

Griffin and coworkers designed an alternative polarization scheme to enhance sensitivity in liquid-state NMR experiments of low-γ spins [167]. They cooled a sample to 90 K and polarized under CE conditions at 140 GHz using a biradical as polarizing agent, and subsequently cross-polarized from ^{1}H to low-γ nuclei, such as ^{13}C and ^{15}N. Instead of dissolving the sample irreversibly, the authors melted it rapidly employing an infrared laser pulse, before acquiring the spectrum in solution. The theoretical enhancement in this case is $\epsilon' = \epsilon T_{\mathrm{obs}}/T_{\mathrm{DNP}}$. Enhancements of $\epsilon = 133\,(\epsilon' = 400)$ were reported for small molecules. The advantage of this concept lies in the fact that the sample is not irreversibly dissolved. It is therefore possible to perform signal averaging; freezing typically requires 60–90 s and the melting <1 s. Compared to D-DNP, the overall enhancements ϵ' are lower because the temperature factor is much smaller for temperature ratios of 300 K/100 K compared to 300 K/1.2 K. Temperature-jump DNP (TJ-DNP) could of course be extended to He cooling, although this would become rather impractical because 1.2 K requires an isolated He bath under reduced pressure, and cooling from 4 K to 1.2 K typically requires 5–10 min in common implementations of D-DNP polarizers. However, the combination of dissolution DNP with laser heating should be considered in order to reduce the volume of the dissolution solvent.

3.4 Polarizing Agents

Polarizing agents represent a core resource for DNP. They must exhibit a long lifetime of an unpaired electron, along with good solubility in the solvent of choice, i.e., in water for most biological applications. The most commonly used polarizing agents are now trityl (triarylmethyl) [44, 168–170] and nitroxide radicals. A thorough list of "older" radicals was compiled by Müller-Warmuth [39], who also characterized some for DNP via the OE mechanism.

Some common trityl radicals are shown in Fig. 17. TEMPO nitroxide radicals are frequently used for O-DNP, but also for SS-DNP. Common forms are 4-hydroxy-TEMPO (TEMPOL, also TEMPO) and 4-oxo-TEMPO (TEMPONE). Han showed that that deuteration and ^{15}N-labelling (^{2}H-^{15}N -TEMPONE) increases the saturation factor for O-DNP [51]. TEMPO-type radicals also facilitate the CE, in particular when used as biradicals. Griffin initially developed BTnE (bis-TEMPO-*n*-ethylene oxide) radicals with variable lengths of the linker [88], and later TOTAPOL (1-(TEMPO-4-oxy)-3-(TEMPO-4-amino)propan-2-ol) [89]. He also explored TEMPO-BDPA biradicals [171], and more recently rigid biradicals bTbk (bis-TEMPO-bisketal) [90]. Theoretical work showed that an orthogonal relative orientation of the electron g-tensors is a crucial requirement for obtaining high enhancement DNP factors [91]. Gafurov et al. explored bisnitroxide radicals for O-DNP but found no advantage over monoradicals [172].

For dissolution DNP polarizing via the SE or TM, radicals with narrow lines are required. This is the case for highly symmetrical trityl radicals, which were originally developed by Nycomed as MRI contrast agents [44, 168, 169]. The most commonly used radical of this kind is OX063 (tris{8-carboxyl-2,2,6,6-tetra[2-(1-hydroxyethyl)]-benzo(1,2-d:4,5-d′)bis(1,3)dithiole-4-yl}methyl sodium salt), shown in Fig. 17. Ardenkjær-Larsen explored trityl biradicals and found lower concentrations for optimal polarization, but at the price of much longer build-up times. Pons explored chlorinated versions of trityl radicals (see

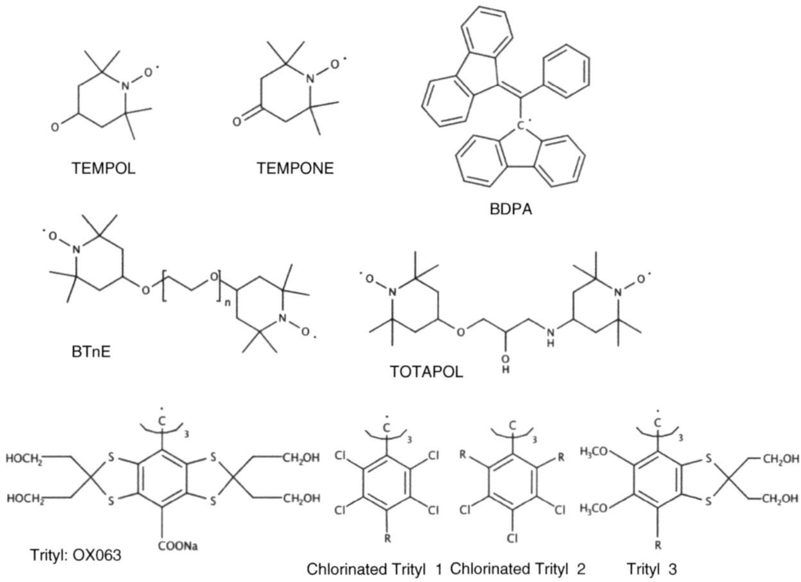

Fig. 17 Chemical structures of radicals used for DNP experiments

Fig. 17), which are also suitable for D-DNP but seem to alter the polarization mechanism, showing marked differences between neutral and anionic forms of the radical [173]. Density functional theory (DFT) calculations support the hypothesis that in these radicals polarization is transferred from the unpaired electron to chlorine nuclei, and from these to carbon by spin diffusion [174].

Recently, Stephan presented a tunable structure that allows easy access to an array of trityl radicals possessing modular surface functional groups (trityl 3 in Fig. 17) [175]. While these radicals are less symmetrical they showed comparable enhancements for small molecules (Stephan and Günther, unpublished results from this laboratory).

Corzilius recently reported high-spin transition metal compounds (Mn^{2+} and Gd^{3+}) as polarizing agents. When the EPR linewidth is narrow and there is no spectral dilution by strong hyperfine coupling to the metal nucleus, DNP performance was shown to be comparable to that obtained with trityl radicals. This finding puts the reports on larger polarizations employing Gd^{3+} into a different perspective [143, 145], and probably requires reconsideration of the polarization mechanism in the presence of Gd^{3+}.

Recently, Dollmann et al. introduced potentially biocompatible spin-labelled heparins, which have remarkable features for O-DNP [176]. For these compounds, the off-resonant EPR hyperfine lines contribute to the total saturation, even in the absence of Heisenberg spin exchange (HSE) and electron spin–nuclear spin relaxation (T_{1ne}). Alternatively spin-labelled hydrogels were proposed [177], which may have unique advantages for shuttle DNP experiments owing to the prolonged lifetime of the hyperpolarization. Münnemann also explored a spin-labelled cationic polyelectrolyte poly(DADMAC) [a copolymer of poly(diallyldi-methyl-ammonium chloride with 4% nitroxide radical-bearing monomers], as mentioned earlier [115].

4 Conclusions

Without doubt, DNP applications have started to open new avenues for increasing the range of NMR applications. However, there is no universal solution that provides large enhancements for a broad range of NMR application. The only implementation directly applicable to biological macromolecules is SS-DNP, although enhancements for proteins seem to be smaller than for small molecules and further work is needed to enable wide-spread applications for biological macromolecules. This is, however, clearly facilitated by substantial groundwork and the availability of commercial implementations.

D-DNP has enormous potential for medical applications using metabolic imaging, and is likely to provide exciting new solutions for bioanalytical NMR. However, the requirement for He cooling considerably increases the operational costs of the spectrometer, and the requirement for sample freezing, dissolution and transfer limits applications. Nevertheless, driven by the immense enhancements for some molecules, a range of interesting solutions has already been proposed. D-DNP also benefits from a commercially available polarizer and emerging solution that speed up sample dissolution and transfer. For polarizations carried out in the solid state, the theoretical description remains incomplete.

O-DNP, the only DNP-method carrying out polarizations in liquids, has benefited from a series of mechanistic studies to characterize the polarization parameters. From this groundwork it is becoming increasingly clear that considerable enhancements are feasible, although sample heating remains a significant problem. Shuttling or flow solutions present an interesting compromise but are limited to modest enhancements, whereas polarization and acquisition at higher field require further optimizations, implementing significant cooling and polarization cells enabling higher resolutions on NMR spectra.

Although this review is mainly focused on DNP, there are several parallel developments generating hyperpolarization sources other than unpaired electrons, some of which were mentioned in the "Introduction". Some methods yield 100% theoretical polarization, although requiring a different set of compromises to realize these enhancements for practical samples.

Without doubt, hyperpolarization methods have entered the arena of biomedical NMR, although more work is needed to provide practical implementations. Considering that the first DNP experiment was carried out almost half a century ago, it is astonishing that many polarization mechanisms remain poorly understood, leaving exiting opportunities for further work.

Acknowledgments I would like to thank Marina Bennati and Thomas Prisner for helpful discussions. I would like to thank Christian Ludwig for providing additional data for Fig. 16d. I also thank Christian Ludwig and Rafael Brüschweiler for critical proof reading.

References

1. Abragam A (1961) Principles of nuclear magnetism. Oxford University Press, London
2. Jeffries CD (1964) Ann Rev Nucl Sci 14:101
3. Bouchiat M, Carver T, Varnum C (1960) Phys Rev Lett 5:373
4. Eisenschmid TC, Kirss RU, Deutsch PP, Hommeltoft SI, Eisenberg R, Bargon J, Lawler RG, Balch AL (1987) J Am Chem Soc 109:8089
5. Bowers CR, Weitekamp DP (1986) Phys Rev Lett 57:2645
6. Bargon J, Fischer H, Johnsen U (1967) Z Naturforsch A 22:1551
7. Ward HR, Lawler RG (1967) J Am Chem Soc 89:5518
8. Closs GL (1969) J Am Chem Soc 91:4552
9. Kaptein R, Dijkstra K, Nicolay K (1978) Nature 274:293
10. Adams R, Aguilar J, Atkinson K, Cowley M, Elliott P, Duckett S, Green G, Khazal I, Lopez-Serrano J, Williamson D (2009) Science 323:1708
11. Haupt J (1972) Phys Lett A A 38:389
12. Horsewill AJ (1999) Prog Nucl Magn Reson Spectrosc 35:359
13. Ludwig C, Saunders M, Marin-Montesinos I, Günther UL (2010) Proc Natl Acad Sci USA 107:10799
14. Tomaselli M, Meier U, Meier BH (2004) J Chem Phys 120:4051
15. Tomaselli M, Knecht DW, Holleman I, Meijer G, Meier BH (2000) J Chem Phys 113:5141
16. Tomaselli M, Meier BH (2001) J Chem Phys 115:11017
17. Bifone A, Song YQ, Seydoux R, Taylor RE, Goodson BM, Pietrass T, Budinger TF, Navon G, Pines A (1996) Proc Natl Acad Sci USA 93:12932
18. Lampel G (1968) Phys Rev Lett 20:491
19. Albert MS, Cates GD, Driehuys B, Happer W, Saam B, Springer CS Jr, Wishnia A (1994) Nature 370:199
20. Navon G, Song Y-Q, Rõõm T, Appelt S, Taylor R, Pines A (1996) Science 271:1848
21. Overhauser AW (1953) Phys Rev 92:411
22. Carver TR, Slichter CP (1953) Phys Rev 92:212
23. Carver TR, Slichter CP (1956) Phys Rev 102:975
24. Pound RV (1950) Phys Rev 79:685
25. Abragam A (1955) Phys Rev 98:1729
26. Jeffries CD (1957) Phys Rev 106:164
27. Borghini M (1968) Phys Rev Lett 20:419
28. Borghini M, Scheffler K (1971) Phys Rev Lett 26:1362
29. Abragam A, Proctor W (1958) C R Hebd Seances Acad Sci 246:2253
30. Kessenikh A, Lushchinov V, Manekov A, Taran Y (1963) Sov Phys Solid State 5:321
31. Kessenikh A, Manenkov A, Lyatnitskii G (1964) Sov Phys Solid State 6:641
32. Hwang CF, Hill D (1967) Phys Rev Lett 18:110
33. Hwang CF, Hill DA (1967) Phys Rev Lett 19:1011
34. Byvik CE, Wollan DS (1974) Phys Rev B 10:791
35. Wollan DS (1976) Phys Rev B 13:3686
36. Wind RA, Duijvestijn MJ, van der Lugt C, Manenschijn A, Vriend J (1985) Prog Nucl Magn Reson Spectrosc 17:33
37. Hausser K, Stehlik D (1968) Adv Magn Reson 3:79
38. Hall D, Maus D, Gerfen G, Inati S, Becerra L, Dahlquist F, Griffin R (1997) Science 276:930

39. Müller-Warmuth W, Meise-Gresch K (1983) Adv Magnn Reson 11:1
40. Bennati M, Tkach I, Türke M-T (2011) Electron Paramagn Reson 22:155
41. Abragam A, Goldman M (1978) Rep Prog Phys 41:395
42. Maly T, Debelouchina GT, Bajaj VS, Hu KN, Joo CG, Mak-Jurkauskas ML, Sirigiri JR, van der Wel PC, Herzfeld J, Temkin RJ, Griffin RG (2008) J Chem Phys 128:052211
43. Hovav Y, Feintuch A, Vega S (2011) J Chem Phys 134:074509
44. Ardenkjaer-Larsen JH, Fridlund B, Gram A, Hansson G, Hansson L, Lerche MH, Servin R, Thaning M, Golman K (2003) Proc Natl Acad Sci 100:10158
45. Slichter CP (2010) Phys Chem Chem Phys 12:5741
46. Denninger G, Stocklein W, Dormann E, Schwoerer M (1984) Chem Phys Lett 107:222
47. Stocklein W, Seidel H, Singel D, Kendrick R, Yannoni C (1987) Chem Phys Lett 141:277
48. Wind RA, Lock H, Mehring M (1987) Chem Phys Lett 141:283
49. Solomon I (1955) Phys Rev Lett 99:959
50. Bates R Jr, Drozdoski W (1977) J Chem Phys 67:4038
51. Armstrong B, Han S (2007) J Chem Phys 127:104508
52. Abragam A, Pound R (1953) Phys Rev 92:943
53. Hwang LP, Freed JH (1975) J Chem Phys 63:4017
54. Bennati M, Luchinat C, Parigi G, Türke M (2010) Phys Chem Chem Phys 12:5902
55. Armstrong BD, Han S (2009) J Am Chem Soc 131:4641
56. Reese M, Türke MT, Tkach I, Parigi G, Luchinat C, Marquardsen T, Tavernier A, Hofer P, Engelke F, Griesinger C, Bennati M (2009) J Am Chem Soc 131:15086
57. Lingwood M, Han S (2009) J Magn Reson 201:137
58. Sezer D, Prandolini MJ, Prisner TF (2009) Phys Chem Chem Phys 11:6626
59. Kryukov EV, Pike KJ, Tam TKY, Newton ME, Smith ME, Dupree R (2011) Phys Chem Chem Phys 13:4372
60. Jeffries CD (1960) Phys Rev 117:1056
61. Wenckebach WT (2008) Appl Magn Reson 34:227
62. Henstra A, Wenckebach WT (2008) Mol Phys 106:859
63. Jeschke G, Schweiger A (1996) Mol Phys 88(2):335
64. Henstra A, Dirksen P, Schmidt J, Wenckebach WT (1988) J Magn Reson 77:389
65. Wind R, Lock H (1990) Adv Magn Opt Reson 15:51
66. Ludwig C, Marin-Montesinos I, Saunders MG, Günther UL (2010) J Am Chem Soc 132:2508
67. Lumata L, Kovacs Z, Malloy C, Sherry AD, Merritt M (2011) Phys Med Biol 56:N85
68. Bloembergen N (1949) Physica 15:3386
69. Dementyev AE, Cory DG, Ramanathan C (2008) Phys Rev Lett 100:127601
70. Saunders MG, Ludwig C, Günther UL (2008) J Am Chem Soc 130:6914
71. Abragam A, Borghini M (1964) In: Gorter C (ed) Progress in low temperature physics, vol IV. North-Holland Publishing Company, Amsterdam
72. Goldman M (1970) Spin temperature and nuclear magnetic resonance in solids. Oxford University Press, Oxford
73. Goertz S (2004) Nucl Instrum Methods Phys Res A 526:28
74. Radtke E, Reicherz G, Schiemann M (2008) Appl Magn Reson 34:461
75. Wind R, Li L, Lock H, Maciel G (1988) J Magn Reson 79:577
76. Henstra A, Dirksen P, Wenckebach WT (1988) Phys Lett A 134:134
77. Henstra A, Lin T, Schmidt J, Wenckebach W (1990) Chem Phys Lett 165:6
78. Weis V, Bennati M, Rosay M, Griffin RG (2000) J Chem Phys 113:6795
79. Morley GW, van Tol J, Ardavan A, Porfyrakis K, Zhang J, Briggs GA (2007) Phys Rev Lett 98:220501
80. Bajaj V, Farrar C, Hornstein M, Mastovsky I, Vieregg J, Bryant J, Elena B, Kreischer K, Temkin R, Griffin R (2003) J Magn Reson 160:85
81. Bajaj VS, Hornstein MK, Kreischer KE, Sirigiri JR, Woskov PP, Mak-Jurkauskas ML, Herzfeld J, Temkin RJ, Griffin RG (2007) J Magn Reson 189:251

82. Maly T, Andreas LB, Smith AA, Griffin RG (2010) Phys Chem Chem Phys 12:5872
83. Wind R, Anthonio F, Duijvestijn M, Smidt J, Trommel J, de Vette G (1983) J Magn Reson 52:424
84. Afeworki M, McKay RA, Schaefer J (1992) Macromolecules 25:4084
85. Becerra LR, Gerfen GJ, Temkin RJ, Singel DJ, Griffin RG (1993) Phys Rev Lett 71:3561
86. Becerra L, Gerfen G, Bellew B, Bryant J, Hall D, Inati S, Weber R, Un S, Prisner T, McDermott A (1995) J Magn Reson Series A 117:28
87. Hu KN, Yu H, Swager TM, Griffin RG (2004) J Am Chem Soc 126:10844
88. Song C, Hu KN, Joo CG, Swager TM, Griffin RG (2006) J Am Chem Soc 128:11385
89. Matsuki Y, Maly T, Ouari O, Karoui H, Le Moigne F, Rizzato E, Lyubenova S, Herzfeld J, Prisner T, Tordo P, Griffin RG (2009) Angew Chem Int Ed Engl 48:4996
90. Ysacco C, Rizzato E, Virolleaud M-A, Karoui H, Rockenbauer A, Le Moigne F, Siri D, Ouari O, Griffin RG, Tordo P (2010) Phys Chem Chem Phys 12:5841
91. Thurber KR, Tycko R (2010) Phys Chem Chem Phys 12:5779
92. Barnes AB, Corzilius B, Mak-Jurkauskas ML, Andreas LB, Bajaj VS, Matsuki Y, Belenky ML, Lugtenburg J, Sirigiri JR, Temkin RJ, Herzfeld J, Griffin RG (2010) Phys Chem Chem Phys 12:5861
93. Debelouchina GT, Bayro MJ, Van Der Wel PCA, Caporini MA, Barnes AB, Rosay M, Maas WE, Griffin RG (2010) Phys Chem Chem Phys 12:5911
94. Akbey U, Franks WT, Linden A, Lange S, Griffin RG, van Rossum BJ, Oschkinat H (2010) Angew Chem Int Ed Engl 49:7803
95. Maly T, Miller AF, Griffin RG (2010) Chemphyschem 11:999
96. Lesage A, Lelli M, Gajan D, Caporini MA, Vitzthum V, Mieville P, Alauzun J, Roussey A, Thieuleux C, Mehdi A, Bodenhausen G, Coperet C, Emsley L (2010) J Am Chem Soc 132:15459
97. Idehara T, Saito T, Ogawa I, Mitsudo S, Tatematsu Y, Agusu L, Mori H, Kobayashi S (2008) Appl Magn Reson 34:265
98. Feintuch A, Shimon D, Hovav Y, Banerjee D, Kaminker I, Lipkin Y, Zibzener K, Epel B, Vega S, Goldfarb D (2011) J Magn Reson 209:136
99. Hunter RI, Cruickshank PAS, Bolton DR, Riedi PC, Smith GM (2010) Phys Chem Chem Phys 12:5752
100. Thurber KR, Yau WM, Tycko R (2010) J Magn Reson 204:303
101. Dorn HC, Wang J, Allen L, Sweeney D, Glass T (1988) J Magn Reson 79:404
102. Stevenson S, Dorn HC (1994) Anal Chem 66:2993
103. Stevenson S, Glass T, Dorn H (1998) Anal Chem 70:2623
104. Dorn H, Gitti R, Tsai K, Glass T (1989) Chem Phys Lett 155:227
105. Sudmeier JL, Günther UL, Albert K, Bachovchin WW (1996) J Magn Reson A 118:145
106. Lingwood M, Siaw T, Sailasuta N, Ross B, Bhattacharya P, Han S (2010) J Magn Reson 205:247
107. Purcell EM, Pound RV (1951) Phys Rev 81:279
108. Redfield AG (2003) Magn Reson Chem 41:753
109. Grosse S, Gubaydullin F, Scheelken H, Vieth HM, Yurkovskaya AV (1999) Appl Magn Reson 17:211
110. Victor K, Kavolius V, Bryant R (2004) J Magn Reson 171:253
111. Noack F (1986) Prog Nucl Magn Reson Spectrosc 18:171
112. Reese M, Lennartz D, Marquardsen T, Höfer P, Tavernier A, Carl P, Schippmann T, Bennati M, Carlomagno T, Engelke F, Griesinger C (2008) Appl Magn Reson 34:301
113. Krahn A, Lottmann P, Marquardsen T, Tavernier A, Türke M, Reese M, Leonov A, Bennati M, Hoefer P, Engelke F (2010) Phys Chem Chem Phys 12:5830
114. Zotev VS, Owens T, Matlashov AN, Savukov IM, Gomez JJ, Espy MA (2010) J Magn Reson 207:78
115. Münnemann K, Bauer C, Schmiedeskam J, Spiess H, Schreiber W, Hiderberger D (2008) Appl Magn Reson 34:321

116. Armstrong B, Lingwood M, McCarney E, Brown E, Blümler P, Han S (2008) J Magn Reson 191:273
117. McDermott R (2002) Science 295:2247
118. Guiberteau T, Grucker D (1996) J Magn Reson Series B 110:47
119. Halse ME, Callaghan PT (2008) J Magn Reson 195:162
120. Türke MT, Tkach I, Reese M, Höfer P, Bennati M (2010) Phys Chem Chem Phys 12:5893
121. Höfer P, Parigi G, Luchinat C, Carl P, Guthausen G, Reese M, Carlomagno T, Griesinger C, Bennati M (2008) J Am Chem Soc 130:3254
122. Villanueva-Garibay JA, Annino G, Van Bentum PJM, Kentgens APM (2010) Phys Chem Chem Phys 12:5846
123. Denysenkov V, Prandolini MJ, Gafurov M, Sezer D, Endeward B, Prisner TF (2010) Phys Chem Chem Phys 12:5786
124. Prandolini MJ, Denysenkov VP, Gafurov M, Endeward B, Prisner TF (2009) J Am Chem Soc 131:6090
125. Sezer D, Gafurov M, Prandolini MJ, Denysenkov VP, Prisner T (2009) Phys Chem Chem Phys 11:6638
126. Kryukov EV, Newton ME, Pike KJ, Bolton DR, Kowalczyk RM, Howes AP, Smith ME, Dupree R (2010) Phys Chem Chem Phys 12:5757
127. Loening N, Rosay M, Weis V, Griffin R (2002) J Am Chem Soc 124:8808
128. Kausik R, Han S (2009) J Am Chem Soc 131:18254
129. Armstrong BD, Choi J, Lopez C, Wesener DA, Hubbell W, Cavagnero S, Han S (2011) J Am Chem Soc 133:5987
130. Sowerby A (2005) Chem Ind:21
131. Ardenkjaer-Larsen JH, Leach AM, Clarke N, Urbahn J, Anderson D, Skloss TW (2011) NMR Biomed 24:927–932
132. Comment A, van den Brandt B, Uffmann K, Kurdzesau F, Jannin S, Konter JA, Hautle P, Wenckebach WT, Gruetter R, Van Der Klink J (2007) Concepts Magn Reson 31B:225
133. Leggett J, Hunter R, Granwehr J, Panek R, Perez-Linde AJ, Horsewill AJ, McMaster J, Smith G, Kockenberger W (2010) Phys Chem Chem Phys 12:5883
134. Bowen S, Hilty C (2010) Phys Chem Chem Phys 12:5766
135. Liu Y, Villamena FA, Rockenbauer A, Rockenbauer A, Zweier J (2010) Chem Commun (Camb) 46:628
136. Day SE, Kettunen MI, Gallagher FA, Hu DE, Lerche M, Wolber J, Golman K, Ardenkjaer-Larsen JH, Brindle KM (2007) Nat Med 13:1382
137. Brindle K (2008) Nat Rev Cancer 8:94
138. Gallagher FA, Kettunen MI, Day SE, Hu D-E, Ardenkjaer-Larsen JH, Zandt RIt, Jensen PR, Karlsson M, Golman K, Lerche MH, Brindle KM (2008) Nature 453:940
139. Nelson S, Vigneron D, Kurhanewicz J, Chen A, Bok R, Hurd D (2008) Appl Magn Reson 34:533
140. Golman K, Zandt RI, Lerche M, Pehrson R, Ardenkjaer-Larsen JH (2006) Cancer Res 66:10855
141. Ardenkjaer-Larsen J, Macholl S, Johannesson H (2008) Appl Magn Reson 34:509
142. Reynolds S, Patel H (2008) Appl Magn Reson 34:495
143. Day IJ, Mitchell JC, Snowden MJ, Davis AL (2007) Magn Reson Chem 45:1018
144. Macholl S, Johannesson H, Ardenkjaer-Larsen JH (2010) Phys Chem Chem Phys 12:5804
145. Johannesson H, Macholl S, Ardenkjaer-Larsen JH (2009) J Magn Reson 197:167
146. Jannin S, Comment A, Kurdzesau F, Konter JA, Hautle P, van den Brandt B, van der Klink JJ (2008) J Chem Phys 128:241102
147. Emwas A-H, Saunders M, Ludwig C, Günther U (2008) Appl Magn Reson 34:483
148. Comment A, Rentsch J, Kurdzesau F, Jannin S, Uffmann K, van Heeswijk RB, Hautle P, Konter JA, van den Brandt B, van der Klink JJ (2008) J Magn Reson 194:152
149. Kurdzesau F, van den Brandt B, Comment A, Hautle P, Jannin S, Van Der Klink J, Konter J (2008) J Phys D: Appl Phys 41:155506

150. Mishkovsky M, Frydman L (2008) Phys Chem Chem Phys 9:2340
151. Frydman L, Blazina D (2007) Nat Phys 3:415
152. Panek R, Granwehr J, Leggett J, Kockenberger W (2010) Phys Chem Chem Phys 12:5771
153. Ludwig C, Marin-Montesinos I, Saunders MG, Emwas AH, Pikramenou Z, Hammond SP, Günther UL (2010) Phys Chem Chem Phys 12:5868
154. Bowen S, Zeng H, Hilty C (2008) Anal Chem 80:5794
155. Zeng H, Bowen S, Hilty C (2009) J Magn Reson 199:159
156. Wilson D, Hurd R, Keshari K et al. (2009) Proc Natl Acad Sci USA 106:5503
157. Vasos PR, Comment A, Sarkar R, Ahuja P, Jannin S, Ansermet JP, Konter JA, Hautle P, van den Brandt B, Bodenhausen G (2009) Proc Natl Acad Sci USA 106:18469
158. Pileio G, Carravetta M, Levitt MH (2010) Proc Natl Acad Sci USA 107:17135
159. Carravetta M, Johannessen OG, Levitt MH (2004) Phys Rev Lett 92:153003
160. Carravetta M, Levitt MH (2004) J Am Chem Soc 126:6228
161. Warren WS, Jenista E, Branca RT, Chen X (2009) Science 323:1711
162. Bowen S, Hilty C (2008) Angew Chem Int Ed Engl 47:5235
163. Harris T, Giraudeau P, Frydman L (2011) Chemistry 17:697
164. Bowen S, Sekar G, Hilty C (2011) NMR Biomed 24:1016–1022
165. Jensen PR, Meier S, Ardenkjaer-Larsen JH, Duus JO, Karlsson M, Lerche MH (2009) Chem Commun (Camb):5168
166. Lerche MH, Meier S, Jensen PR, Baumann H, Petersen BO, Karlsson M, Duus JO, Ardenkjaer-Larsen JH (2010) J Magn Reson 203:52
167. Joo CG, Hu KN, Bryant JA, Griffin RG (2006) J Am Chem Soc 128:9428
168. Ardenkjaer-Larsen JH, Laursen I, Leunbach I, Ehnholm G, Wistrand LG, Petersson JS, Golman K (1998) J Magn Reson 133:1
169. Thaning M (2000) US Patent 6,013,810
170. Wind RA, Ardenkjaer-Larsen JH (1999) J Magn Reson 141:347
171. Dane EL, Maly T, Debelouchina GT, Griffin RG, Swager TM (2009) Org Lett 11:1871
172. Gafurov M, Lyubenova S, Denysenkov V, Ouari O, Karoui H, Le Moigne F, Tordo P, Prisner T (2010) Appl Magn Reson 37:505
173. Gabellieri C, Mugnaini V, Paniagua JC, Roques N, Oliveros M, Feliz M, Veciana J, Pons M (2010) Angew Chem Int Ed Engl 49:3360
174. Paniagua JC, Mugnaini V, Gabellieri C, Feliz M, Roques N, Veciana J, Pons M (2010) Phys Chem Chem Phys 12:5824
175. Stephan M, Günther UL (PhosPhoenix SARL) Patent FR2010 04364
176. Dollmann BC, Kleschyov AL, Sen V, Golubev V, Schreiber L, Spiess H, Münnemann K, Hinderberger D (2010) Chemphyschem 11:3656
177. Dollmann BC, Junk MJN, Drechsler M, Spiess HW, Hinderberger D, Münnemann K (2010) Phys Chem Chem Phys 12:5879
178. Pavlovskaya GE, Cleveland ZI, Stupic KF, Basaraba RJ, Meersmann T (2005) Proceedings of the National Academy of Sciences of the United States of America 102:18275
179. Türke M-T, Bennati M (2011) Phys Chem Chem Phys 13:3630–3633
180. Barnes A, De Paëpe G, Van Der Wel P, Hu K, Joo C, Bajaj V, Mak-Jurkauskas M, Sirigiri J, Herzfeld J, Temkin R, Griffin R (2008) Appl Magn Reson 34:237

Top Curr Chem (2013) 335: 71–96
DOI: 10.1007/128_2011_226

Published online: 12 August 2011

NMR with Multiple Receivers

Ēriks Kupče

Abstract Parallel acquisition NMR spectroscopy (PANSY) is used to detect simultaneously signals from up to four nuclear species, such as H-1, H-2, C-13, N-15, F-19 and P-31. The conventional COSY, TOCSY, HSQC, HMQC and HMBC pulse sequences have been adapted for such applications. Routine availability of NMR systems that incorporate multiple receivers has led to development of new types of NMR experiments. One such scheme named PANACEA allows unambiguous structure determination of small organic molecules from a single measurement and includes an internal field/frequency correction routine. It does not require the conventional NMR lock system and can be recorded in pure liquids. Furthermore, long-range spin–spin couplings can be extracted from the PANACEA spectra and used for three-dimensional structure refinement. In bio-molecular NMR, multi-receiver NMR systems are used for simultaneous recording of H-1 and C-13 detected multi-dimensional spectra. For instance, the 2D (HA)CACO and 3D (HA)CA(CO)NNH experiments can be recorded simultaneously in proteins of moderate size (up to 30 kDa). The multi-receiver experiments can also be used in combination with the fast acquisition schemes such as Hadamard spectroscopy, computer optimized aliasing and projection-reconstruction techniques. In general, experiments that utilize multiple receivers provide significantly more information from a single NMR measurement as compared to the conventional single receiver techniques.

Keywords Hadamard spectroscopy · Multidimensional NMR · Multiple receivers · Parallel acquisition NMR · Projection reconstruction

---

Ē. Kupče
Agilent Technologies, NMR and MRI Systems, Yarnton, Oxford OX5 1QU, UK
e-mail: eriks.kupce@agilent.com

Contents

Abbreviations

COSY	Correlation spectroscopy
FID	Free induction decay
HETCOR	Heteronuclear correlation
HMBC	Heteronuclear multiple-bond correlation
HMQC	Heteronuclear multiple-quantum correlation
HSQC	Heteronuclear single-quantum correlation
INADEQUATE	Incredible natural abundance double quantum transfer experiment
IPAP	In-phase anti-phase
NMR	Nuclear magnetic resonance
NOE	Nuclear Overhauser effect
PANACEA	Parallel acquisition NMR and all-in-one combination of experimental applications
PANSY	Parallel acquisition NMR spectroscopy
PR	Projection reconstruction
RF	Radio frequency
S/N	Signal-to-noise
TOCSY	Total correlation spectroscopy
TROSY	Transverse relaxation optimized spectroscopy

1 Introduction

NMR technology has always progressed along three main avenues – development of NMR magnets, probes and electronics (NMR consoles). A constant push for increased magnetic field strength is driven by the advantages offered by increased resolution and sensitivity, which in turn allow studies of increasingly complex objects, such as bio-molecules. On the other hand, continuing progress in NMR probe development has provided a further increase in the sensitivity of NMR experiments, particularly with the advent of cryogenic probe technology. The diversity of probes has also increased dramatically over the last two decades. Finally, the data quality and experimental capabilities of the state-of-the-art consoles is also steadily improving. Multiple channels, digital receivers, high speed digitizers and, more recently, multiple receivers are just a few of the most important improvements.

One of the consequences of all these developments is a significant improvement in sensitivity of NMR experiments. As a result the modern NMR experiments are often sampling limited rather than sensitivity limited. This has triggered development of the so-called fast NMR techniques [1–21] that considerably reduce the experiment time and increase the sample throughput. The argument is sometimes turned around by claiming that more accurate and better resolved data can be recorded in the same acquisition time. On the other hand, the continuously growing complexity of NMR probes has significantly broadened the scope of NMR experiments. Introduction of multiple RF coils and gradients have considerably increased the diversity of NMR techniques and in particular have stimulated development of multi-dimensional and multi-nuclear NMR spectroscopy [22, 23]. Finally, all that would not be possible without a concurrent development of NMR consoles which these days can house as many as five independent RF channels operating at different NMR frequencies and complemented with triple axis gradient capabilities. All radiofrequency (RF) and gradient channels can be independently amplitude and (RF channels) phase modulated to produce shaped pulses and gradient waveforms. A recent addition to all these technological and electronic marvels is multiple NMR receivers.

Traditionally the NMR spectrometers have been equipped with a single receiver channel and until recently all NMR experiments were designed with a single receiver in mind, often focussing on detection of the most sensitive nuclear species, typically H-1 or F-19. A few attempts to detect simultaneously several nuclear species in the early 1990s were simple one-dimensional experiments that did not really find any widespread applications [24, 25]. On the other hand, the spectacular success of the parallel imaging experiments in MRI [26] did not go unnoticed and nowadays multiple receivers is one of the standard options on MRI systems. The significant difference is, of course, that multiple receiver MRI systems usually detect the same nuclear species, typically H-1 signals received from several (surface) RF coils. Once the multiple receiver technology was introduced and established in MRI, it did not take long before multiple receivers also became available on NMR systems [27]. Up to four receivers can be fitted on the modern NMR consoles and four different nuclear species can be observed simultaneously (see Fig. 1).

This chapter will focus on multi-dimensional NMR experiments that have been designed with multiple receivers in mind. We shall discuss the benefits and challenges of such experiments, their applications in NMR of both small organic molecules and bio-molecules (proteins).

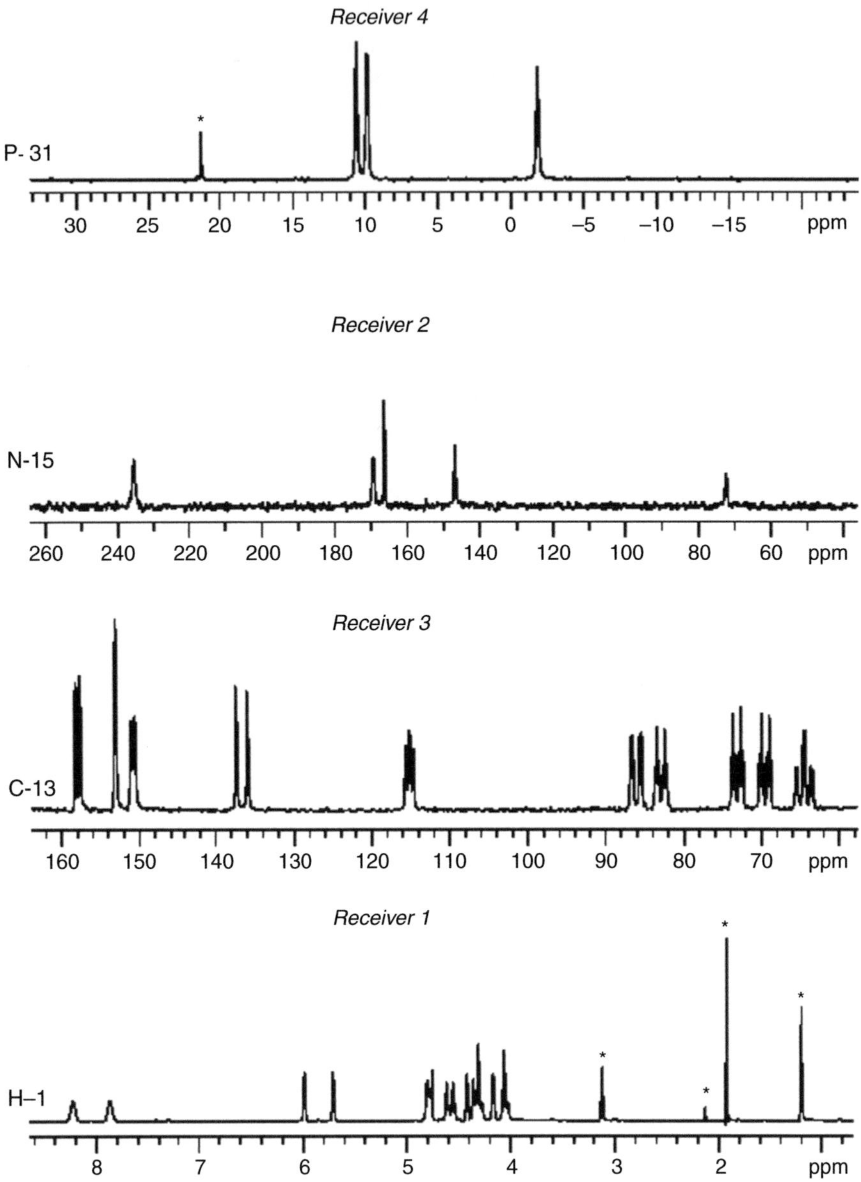

Fig. 1 Spectra of CN labelled guanosine triphosphate recorded simultaneously on a Varian 600 MHz NMR system equipped with four receivers. The peaks from impurities are marked by *asterisks*

2 Basic Two-Dimensional Multi-Receiver Experiments: Parallel vs Sequential Data Acquisition

The classical scheme [28] for a 2D experiment consists of:

Preparation – Evolution – Mixing – Acquisition

On a system equipped with multiple RF channels and receivers several such schemes can be executed in parallel. Let's consider one of the simplest NMR experiments – the two-dimensional COSY experiment. The basic homonuclear COSY pulse sequence consists of an excitation pulse followed by the evolution period, t_1, a read pulse and an acquisition period, t_2 (see Fig. 2a). The same scheme can be executed in parallel on two or more RF channels (Fig. 2b). If the cross-talk between the different nuclear species could be avoided, such an experiment would produce two independent 2D COSY spectra. However, in practise the magnetically active and in particular spin 1/2 nuclei from the same molecule are usually coupled via the scalar spin–spin couplings and in such a simple pulse scheme cross-talk is unavoidable. Therefore we should also observe heteronuclear correlations arising from coherence transfer A $\rightarrow$ X and X $\rightarrow$ A, i.e. in total four two-dimensional spectra in a single measurement! Indeed, all the correlations can be observed in instances where the gyromagnetic ratios of the nuclei and their natural abundances are similar. In fact, as a consequence of the close proximity of the H-1 and F-19 resonance frequencies at the Earth magnetic field, their two-dimensional correlation spectra can be observed even with a single receiver [29].

However, by far more common situations involve nuclear species with rather different gyromagnetic ratios and consequently different NMR sensitivities. As a result, the correlations originating from lower gamma nuclei are considerably weaker as compared to those originating on higher gamma nuclear species (see Fig. 3). Note also that unwanted overlap between the homonuclear and heteronuclear correlations in such experiments can be avoided by shifting the carrier frequencies off-resonance

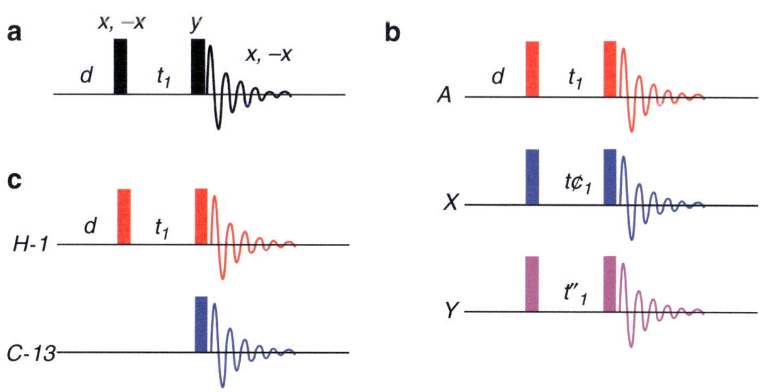

Fig. 2 (**a**) Conventional COSY experiment. (**b**) Super-COSY experiment incorporating several conventional sequences (**a**) applied in parallel on different RF channels. (**c**) PANSY–COSY experiment

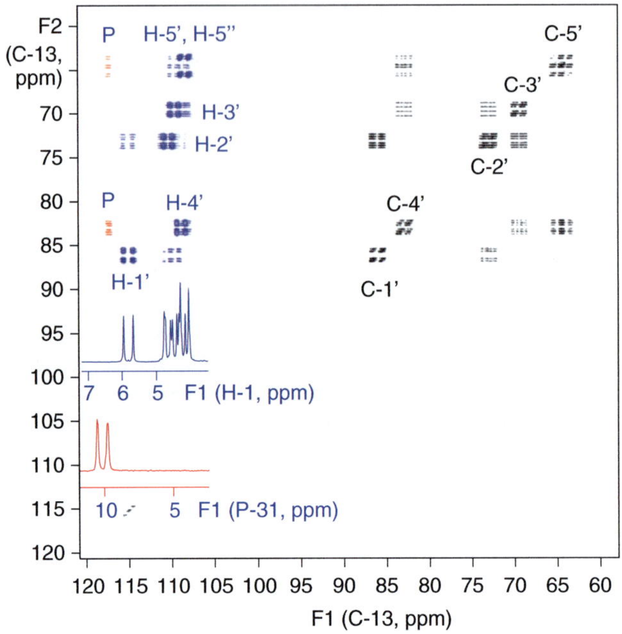

Fig. 3 The C-13 plane of a quadruple COSY spectrum of CN labelled guanosine triphosphate recorded on a Varian 600 MHz NMR system equipped with four receivers. See the pulse sequence in Fig. 2b. Because of the low S/N ratio the correlations involving the N-15 nuclei are not visible in this plane

or by optimizing the offsets to take advantage of the empty spectral regions in the corresponding one-dimensional spectra (projections).

Alternatively, we may consider a heteronuclear COSY experiment (HETCOR) shown in Fig. 2c. It begins in the same way as the homonuclear COSY sequence shown in Fig. 2a, except the read pulse followed by acquisition is usually applied to the X-nuclei. Of course, on a multi-receiver system we may choose to detect signals from both the A and X nuclei (see Fig. 2c). With a proper phase cycle in place and with A = ^{1}H and X = ^{13}C, both ^{1}H–^{1}H and ^{1}H–^{13}C correlation spectra can be recorded simultaneously (see Fig. 4). Note that the X–X correlations and the X $\rightarrow$ A coherence transfer (and therefore any overlap of cross-peaks from different sub-spectra) are avoided in this experiment. This type of experiment is called PANSY (Parallel Acquisition NMR SpectroscopY) [27]. Interestingly, the two-dimensional H–C PANSY–COSY experiment provides rather complete information about small organic molecules – one-bond H–C correlations, long-range H–C correlations, H–H correlations and H–H and H–C scalar coupling constants can all be measured from such spectra. This goes a long way towards establishing the structure of such molecules from just one measurement.

The simple two-dimensional correlation experiments described above use simultaneous acquisition at several NMR frequencies. One drawback of such parallel

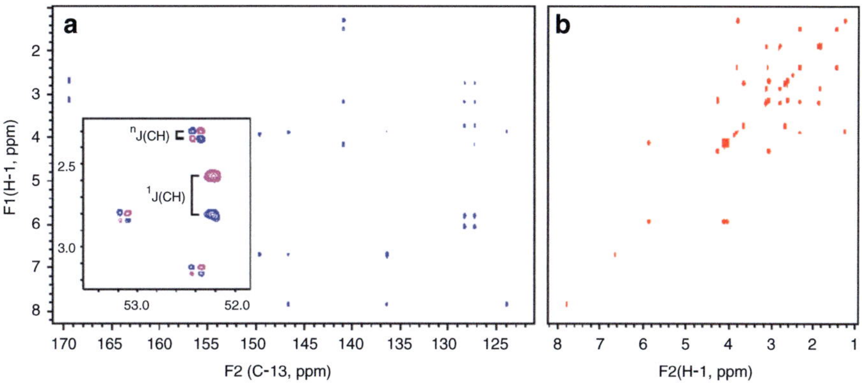

Fig. 4 Two-dimensional PANSY–COSY spectra of brucine recorded simultaneously in ca. 40 min on a Varian 600 MHz NMR system equipped with two receivers. (**a**) H–C COSY spectrum and (**b**) H–H COSY spectrum; 128 complex increments and 4 scans per increment were used. Reproduced from Ref. 27. © The American Chemical Society, 2006

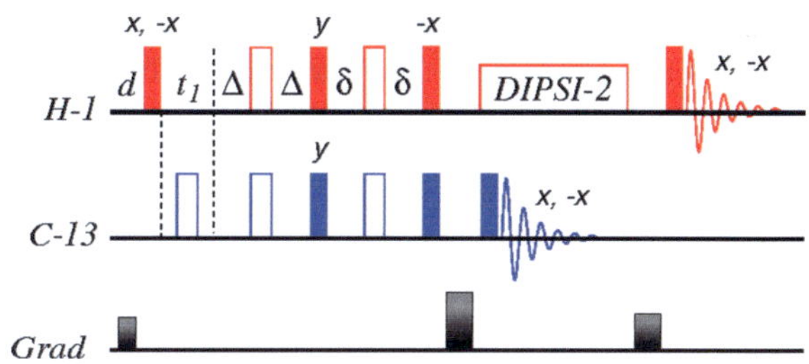

Fig. 5 PANSY–TOCSY pulse sequence for simultaneous acquisition of two-dimensional H–C HETCOR and H–H TOCSY spectra

acquisition is that is does not provide any easy means for decoupling the A and X nuclei, meaning that the corresponding 2D spectra will show A–X splitting in the acquisition dimension (see the inset in Fig. 4a). Fortunately, in dilute spin systems, such as natural abundance C-13 nuclei in small organic molecules, the observed ^{1}H–^{1}H and ^{1}H–^{13}C correlations originate mostly from different molecules and the ^{1}H–^{13}C J-splittings are only present in the C-13 detected spectra. Not only do J-splittings reduce the intensity of the correlation peaks and make the spectra more crowded and difficult to analyse, but such splittings often introduce severe overlap problems. Therefore decoupling A and X spins is generally desirable. This is easily achieved by displacing the A and X acquisition periods. The 2D H–C PANSY TOCSY experiment shown in Fig. 5 does just that.

The experiment starts with an INEPT transfer from H-1 to C-13 nuclei that incorporates t_1 evolution period for H-1 nuclei. The C-13 inversion pulse in the middle of the t_1 period serves to decouple H-1 and C-13 nuclei in the usual manner and has no effect on the H–H correlation spectra because at the natural abundance of isotopes the two spin systems are well separated. The residual H-1 magnetization from C-13 depleted molecules is stored along the *Z*-axis and is subjected to H–H mixing sequence, which at the same time serves to decouple protons from C-13 nuclei. A pair of gradients flanking the mixing/decoupling period is used to purge any unwanted magnetization and requires a pair of C-13 pulses to keep the useful C-13 magnetization along the Z-axis during the first of the two gradient pulses. At this point it is convenient to record the C-13 FID. Thus the H–C correlation experiment is completed by a C-13 read pulse and C-13 data acquisition. The HETCOR spectrum of brucine recorded using the PANSY–TOCSY pulse sequence is shown in Fig. 6a. The two-dimensional H–C correlation spectrum is now decoupled in both, F_1 and F_2 dimensions. Clearly, the resolution in the directly detected C-13 dimension is limited by the duration of the acquisition (i.e. mixing) time. However, because of the decoupling and the rather large chemical shift dispersion in the C-13 domain, this limitation is rather insignificant.

Following the C-13 acquisition period the PANSY–TOCSY experiment continues with a gradient purge pulse followed by a proton read pulse that brings the H-1 magnetization into the transverse plane for the subsequent acquisition of the proton FID. This completes the H–H TOCSY part of the experiment. The two-dimensional H–H TOCSY spectrum of brucine recorded in this way is shown in Fig. 6b. Thus both the H-1 decoupled H–C correlation spectrum and the H–H TOCSY spectrum are recorded simultaneously. Of course, the H-1 and C-13 detected experiments have very different sensitivities and the PANSY spectra recorded as described so far may only be advantageous if the least sensitive of the two experiments is sampling limited rather than sensitivity limited. Similar PANSY experiments involving the somewhat more abundant Si-29 nuclei (4.7%) have been employed to study unstable organosilicon compounds [30].

The experiments described so far serve as simple examples showing that in order to achieve the desired optimum outcome in multi-receiver experiments the corresponding pulse sequences essentially need to be completely re-designed.

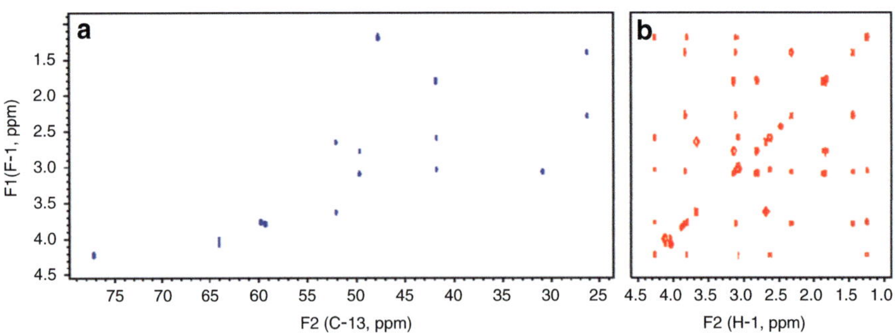

Fig. 6 Two-dimensional PANSY–TOCSY spectra of brucine recorded in ca. 20 min on a Varian 600 MHz NMR system equipped with two receivers. (**a**) H–C HETCOR spectrum and (**b**) H–H TOCSY spectrum; 122 ms long mixing period, 128 complex increments and 2 scans per increment were used. Reproduced from Ref. 27. © The American Chemical Society, 2006

3 Experiments with Parallel Acquisition of H-1 and F-19 FID-s

Both H-1 and F-19 nuclei have comparable NMR sensitivity and therefore PANSY experiments with simultaneous H-1 and F-19 detection seem to be a very attractive option. In fact, at very low magnetic field strength the resonance frequencies of both nuclear species are so close that both resonances can be observed even with a single receiver. For instance, Callaghan et al. have recorded 2D H–F COSY spectra of trifluoroethanol and *para*-difluorobenzene in the Earth's magnetic field [29]. In conventional high-field NMR spectroscopy multi-receiver experiments have been designed to allow simultaneous recording of two-dimensional HSQC, HMQC and HMBC H–X and F–X correlation experiments [31].

The H,F–X HSQC pulse sequence is shown in Fig. 7. Apart from cloning the receiver channel, the necessary adjustments to the conventional HSQC scheme are rather minor. The one-bond F–X couplings are usually considerably larger as compared to the H–X and therefore the F–X polarization transfer delays δ' need be shorter as compared to δ. At the H/F $\rightarrow$ X polarization stage this is conveniently implemented by centring the polarization delays around the simultaneous J-refocusing pulses. The X $\rightarrow$ H/F back-transfer step requires the last F-19 refocusing block be displaced so that both H-1 and F-19 magnetizations are refocused at the beginning of the corresponding acquisition periods. The H–C and F–C HSQC spectra recorded using the described pulse sequence are shown in Fig. 8.

The multiple bond correlation experiments nowadays usually employ coherence transfer gradients because such experiments produce substantially better suppression of unwanted coherences originating mostly from the dominant C-12 molecular sites. A slight complication arises because of the different gyromagnetic ratios of H-1 and F-19 nuclei. The H/F–X HMBC pulse sequence is shown in Fig. 9. Just like with one-bond correlation experiments, the F-19 excitation pulse is delayed to account for generally larger F–X couplings as compared to the H–X couplings. The J-filter designed to suppress correlations due to the one bond H/F–X couplings consists of a 90 X-pulse flanked by bi-polar spoiler gradients, G_s. The encoding gradients G_e are applied during the t_1 evolution period. The encoded magnetization is then decoded by

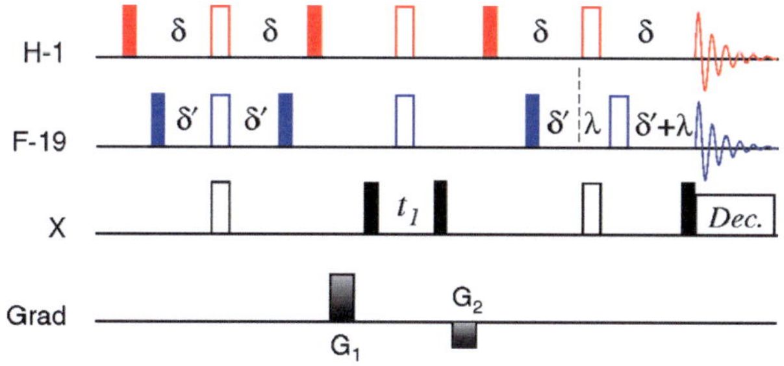

Fig. 7 PANSY–HSQC pulse sequence for parallel acquisition of H–X and F–X correlated spectra. *Filled rectangles* are 90° pulses while *open rectangles* are 180° pulses. Spoiler gradients are denoted G_1 and G_2

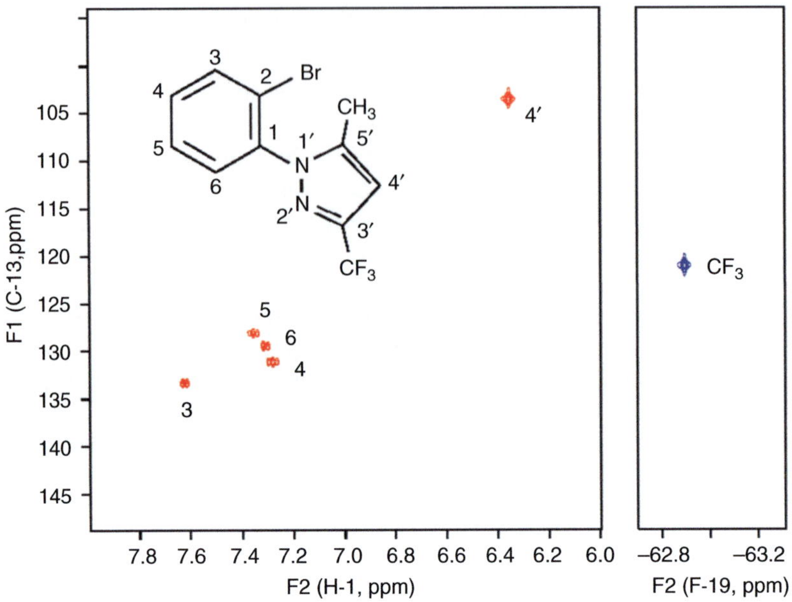

Fig. 8 The HF–X PANSY–HSQC spectra of 2-bromophenyl-3-trifluoromethyl-5-methylpyrazole recorded on a Varian 600 MHz NMR system equipped with four receivers

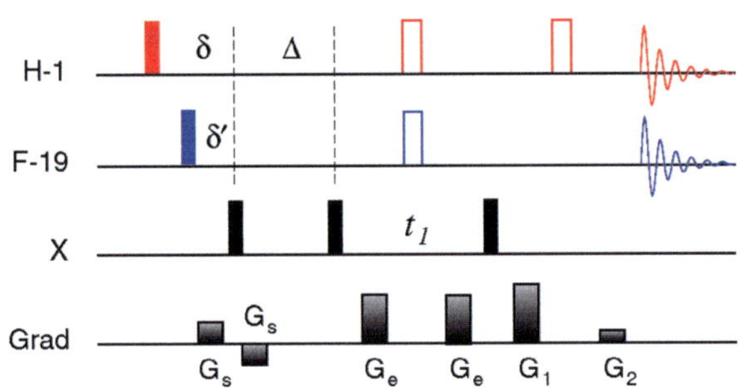

Fig. 9 PANSY–HMBC pulse sequence for simultaneous H–X and F–X correlation experiments. Spoiler gradients are denoted G_s, coherence encoding gradients are denoted G_e. Gradients G_1 and G_2 are coherence decoding gradients

gradients G_1 and G_2. The 180 H-1 pulse takes care of the differences in the gyromagnetic ratios of H-1 and F-19 nuclei. Namely, the F-19 nuclei experience the sum of G_1 and G_2 gradients, while the total effect on the H-1 nuclei is equivalent to the difference of the two gradients, G_1–G_2. The H–C and F–C HMBC spectra recorded simultaneously using this pulse sequence are shown in Fig. 10. Similar experiments have been implemented for X = ^{15}N. Due to lower natural abundance of N-15 nuclei the H–N and F–N HMBC spectra typically take a considerably longer time to record and simultaneous acquisition offers even more considerable time savings.

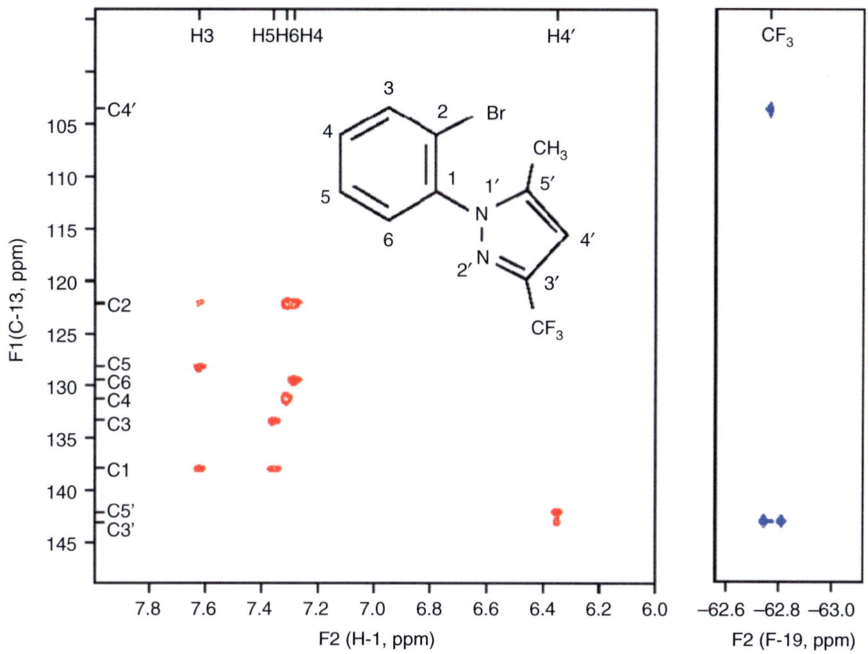

Fig. 10 H/F–C PANSY–HMBC spectra of 2-bromophenyl-3-trifluoromethyl-5-methylpyrazole recorded on a Varian 600 MHz NMR system equipped with four receivers. For clarity reasons only the aromatic region is shown

Finally, it should be noted that in molecular systems with large H–F couplings and complex spin multiplets the H–F decoupling may be increasingly desirable and more sophisticated experiments involving sequential acquisition may be required for such systems.

4 Molecular Structure from a Single Measurement: PANACEA

With recent advances in NMR probe design, sensitivity has become less of a problem and low sensitivity experiments, such as 2D INADEQUATE, are becoming routine. Indeed, the latest generation cryogenic probes allow recording INADEQUATE [32, 33] spectra in as little as 10–15 mg of sample within just a few hours of measurement time. On the other hand, as discussed above, multi-receiver experiments provide new ways of obtaining more information from a single measurement.

Typically the structure of small organic molecules is deduced from a series of measurements. Experiments such as HSQC, HMBC, TOCSY, COSY and TOCSY–HCQC [22] are recorded *sequentially* and the structure of the sample is pieced together from the recorded data once the measurements are finished. Often a set of structures is produced with a corresponding probability factors. Given the technological advances discussed above, it is now possible to design experimental schemes that produce sufficient data for complete structure determination in a single measurement. In fact, the structure determination can be achieved during the experiment and the experiment is then stopped whenever there is sufficient information for the structure determination. One such experiment, named PANACEA (parallel acquisition NMR and all-in-one combination of experimental applications) does just that [34–36].

The multi-receiver PANACEA experiment is designed around the INADEQUATE pulse sequence [32, 33] and also incorporates 1D spectra, multiplicity edited 2D HSQC-s and 3D J-HMBC experiments. In addition, PANACEA does not require the conventional field/frequency lock and can be recorded in pure liquids.

The basic PANACEA pulse sequence is shown in Fig. 11. It starts with the 2D INADEQUATE element. The latter is very similar to the conventional INADEQUATE pulse sequence. First, the C-13 polarization is enhanced by nuclear Overhauser effect generated by proton broad-band decoupling. Then the double-quantum coherence is generated and encoded in the usual way. This is followed by a read pulse and the t_2 acquisition period. At this point the C-13 FID is recorded. A significant difference compared to the conventional INADEQUATE experiment is that the single quantum (SQ) coherence that during the t_1-period is kept along the Z-axis is not suppressed by the phase cycling but rather is saved and later separated from the DQ coherences by post-processing. In this way a one-dimensional C-13 spectrum is also recorded and is then used for several purposes. First, the number of C-13 atoms in the molecule is determined from the spectrum. Second, the position of the SQ resonances is used as a starting point for the INADEQUATE symmetrization procedure that provides an improvement in S/N ratio by a factor of two. Third, co-adding the C-13 traces from the complete data set provides a spectrum with a high S/N ratio. This is useful for detecting any weak signals such as impurities or any minor components of interest. Finally, the individual C-13 spectra can be used for frequency correction and thus the experiment can be recorded without the conventional lock. This allows recording PANACEA spectra in pure liquids, which in turn can be used for fast data acquisition, as discussed below.

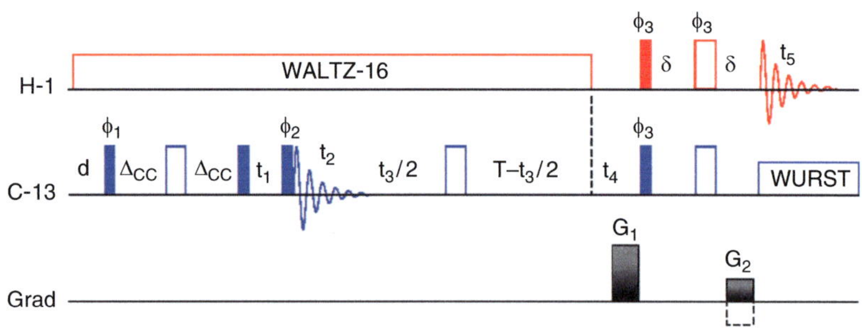

Fig. 11 The basic PANACEA pulse sequence. G_1 and G_2 are coherence transfer gradients. Reproduced from Ref. 34. © The American Chemical Society, 2008

Note that ca. 99% of the C-13 magnetization is in the form of SQ coherence which in the conventional INADEQUATE experiment is suppressed by phase cycling. In the PANACEA experiment the SQ coherence is refocused and transferred to protons for detection (see Fig. 11). The C-13 evolution period, t_3 and the J (CH) evolution period, t_4 are incorporated into the refocusing period, thus reducing sensitivity losses. The C-13 coherences evolve during t_3 in a constant time manner encoding C-13 frequencies for the subsequent multiplicity edited HSQC and three-dimensional HMBC segments. The H-1 decoupling is then switched off and the J (H–C) couplings are allowed to evolve during t_4. The coherence encoding and decoding gradients G_1 and G_2 ensure good suppression of unwanted coherences and clean spectra.

At this point it is useful to remind the reader that the bottleneck of the PANACEA experiment is the low sensitivity INADEQUATE step. Thus, several of the more sensitive H-1 detected spectra can be recorded while the required sensitivity for the INADEQUATE spectrum is achieved. The H-1 detected experiments start with recording of multiplicity edited HSQC spectra. The first three t_4 increments are chosen to be of duration 1/4 J(CH), 2/4 J(CH) and 3/4 J(CH). Thus the first 2D HSQC spectrum contains all CH, CH_2 and CH_3 correlations, the second HSQC spectrum contains only CH correlations and the third HSQC spectrum contains positive CH and CH_3 peaks while the CH_2 peaks appear negative.

Once the edited HSQC experiments are recorded the C-13 decoupling during the H-1 acquisition period t_5 is switched off and the experiment continues as a three-dimensional J-HMBC sequence. This particular setup is designed to ensure that no long-range correlations are lost because of an unfortunate choice of the J(CH) evolution period, t_4. The first three data points that are lost because of the HSQC experiments are recovered by linear prediction. A short version of this experiment may record a 2D HMBC experiment with only one value of the J-evolution delay, t_4 and perhaps one HSQC experiment with $t_4 = 3/4$ J, since the CH and CH_3 groups can easily be distinguished based on the peak intensity information. This of course presumes that the sensitivity of such a short experiment is sufficiently high to record the INADEQUATE spectrum at the same period of time.

The spectra recorded during the basic PANACEA experiment are schematically shown in Fig. 12. The structure of small organic molecules can be established rather unambiguously in five steps: (1) the number of carbon atoms is determined from the one-dimensional C-13 spectrum, (2) the INADEQUATE spectrum provides the connectivity information and established the carbon skeleton of the molecule, (3) the multiplicity edited HSQC spectra provide information about the number of the attached hydrogen atoms, (4) the long range H–C correlations provide the missing connections via any heteroatoms that may be present in the molecule and finally (5) the nature of the heteroatoms is established from the chemical shifts, coupling constants and/or elementary analysis data.

It should be noted that there is plenty of complimentary information in the PANACEA data. For instance, if the C–C correlation is lost in the INADEQUATE spectra because of the degeneracy of the chemical shifts of the adjacent carbon atoms, the missing connectivity can be recovered from the long range H–C

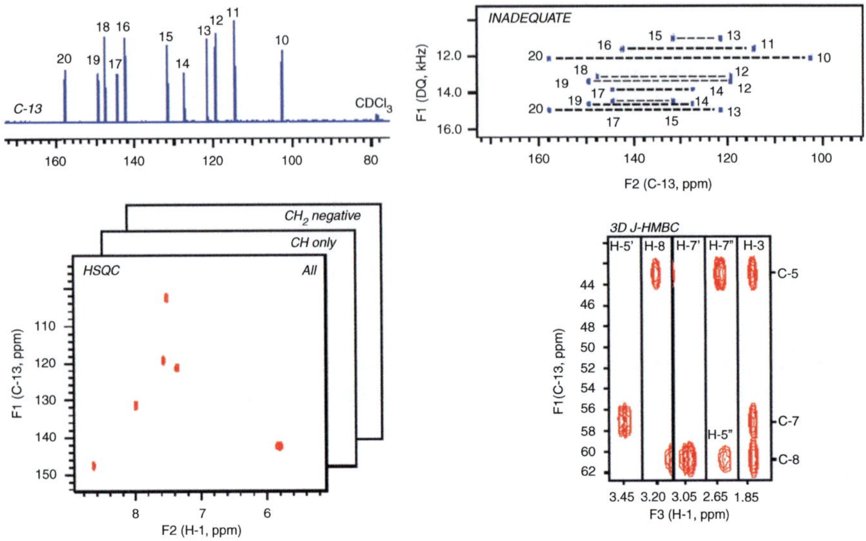

Fig. 12 Selected regions of the PANACEA spectra of quinine recorded in a single experiment using the pulse sequence of Fig. 11. C-13 spectrum, multiplicity edited HSQC spectra and three-dimensional J-HMBC spectrum are all recorded in parallel with the C–C INADEQUATE spectrum. The experiment was recorded on a Varian 600 MHz NMR system equipped with two receivers

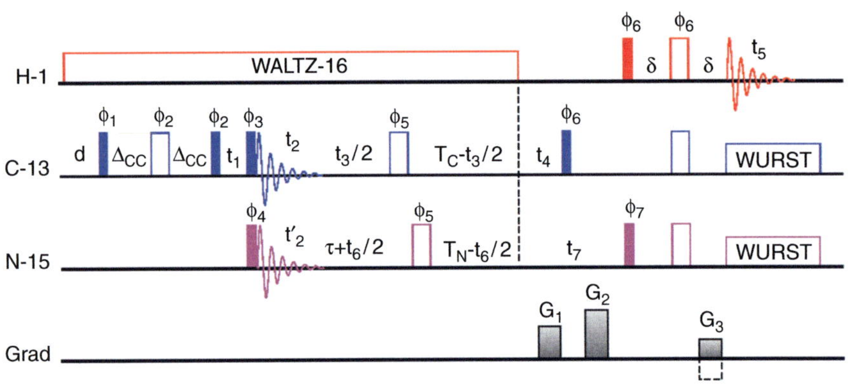

Fig. 13 The extended PANACEA pulse sequence that includes N-15 channel and coherence transfer gradients G_1, G_2 and G_3. Reproduced from Ref. 34.

correlation spectra. Likewise, the multiplicity data are present in both the HSQC spectra and the F_2/F_3 planes of the J-HMBC experiment. Furthermore, the spin–spin coupling constants provide further information about the nature of the C–C and C–H bonds.

For nitrogen containing molecules the PANACEA experiment can be extended to include N-15 detection (see Fig. 13) [34]. Incorporating the N–N INADEQUATE step would be rather impractical for several reasons: (1) the natural abundance of the N-15 isotopes (0.364%) is even lower than that of C-13 (1.07%), (2) cross-talk between the C–C and N–N correlation experiments would have to be suppressed and (3) in the natural products the N–N bonds are by far less common. Hence, the N-15 part of the experiment starts with an N-15 read pulse and acquisition of the N-15 FID. A one-dimensional N-15 spectrum is recorded at this stage. The rest of the N-15 part of the experiment essentially clones the events on the C-13 channel with minor adjustments to accommodate slightly smaller J(NH) coupling constants. An extra gradient pulse (G_2) is inserted to adjust for the differences in the magnitudes of the gyromagnetic ratios of N-15 and C-13. Thus the extended PANACEA experiment provides further multiplicity edited N-15 HSQC spectra and three-dimensional N-15 J-HMBC spectra in the same measurement. Eleven spectra are recorded in such an experiment – one-dimensional C-13 and N-15 spectra, the C–C INADEQUATE, six multiplicity edited N–H and C–H correlated HSQC spectra and two three-dimensional C–H and N–H correlated J-HMBC spectra. Furthermore, the structure of small organic molecules could potentially be determined automatically.

5 Multi-Nuclear Field/Frequency Lock

The individual one-dimensional C-13 traces that are recorded during the PANACEA experiment have found another useful application – these spectra can be used as internal lock ("i-Lock") [36, 37] to correct for frequency (field) instabilities in the course of the rather long data acquisition. This is shown in Fig. 14. For demonstration purposes the conventional lock and the temperature regulation were turned off during this experiment. There are severe frequency variations that are observed in the t_1 time domain (Fig. 14a). The conventional processing then produces essentially unusable PANACEA data as can be appreciated from one of the HMBC planes shown in Fig. 14b. The frequency fluctuations introduced by both the field instabilities and the temperature fluctuations are corrected using the one-dimensional C-13 spectra (Fig. 14c). The conventional processing of the corrected data produces HMBC spectra that are essentially artefact free (see Fig. 14d). Thus the PANACEA experiment can be recorded without the traditional deuterium lock. In fact, any nucleus can be used for field/frequency lock provided a reasonably strong signal can be recorded. For instance, the PANACEA experiments with silicon oil use Si-29 spectra for i-Lock. One of the benefits is that the PANACEA spectra can be recorded in pure liquids, such as silicon oil [30], peanut oil [36] and other natural or synthetic liquids. With the state-of-the-art cryogenic probes the PANACEA experiment in such samples becomes sampling limited, rather than sensitivity limited, and the fast NMR techniques can be applied in order to reduce the time that is typically needed for the PANACEA experiments. In fact, the i-Lock approach can easily be adapted for recording many other conventional experiments.

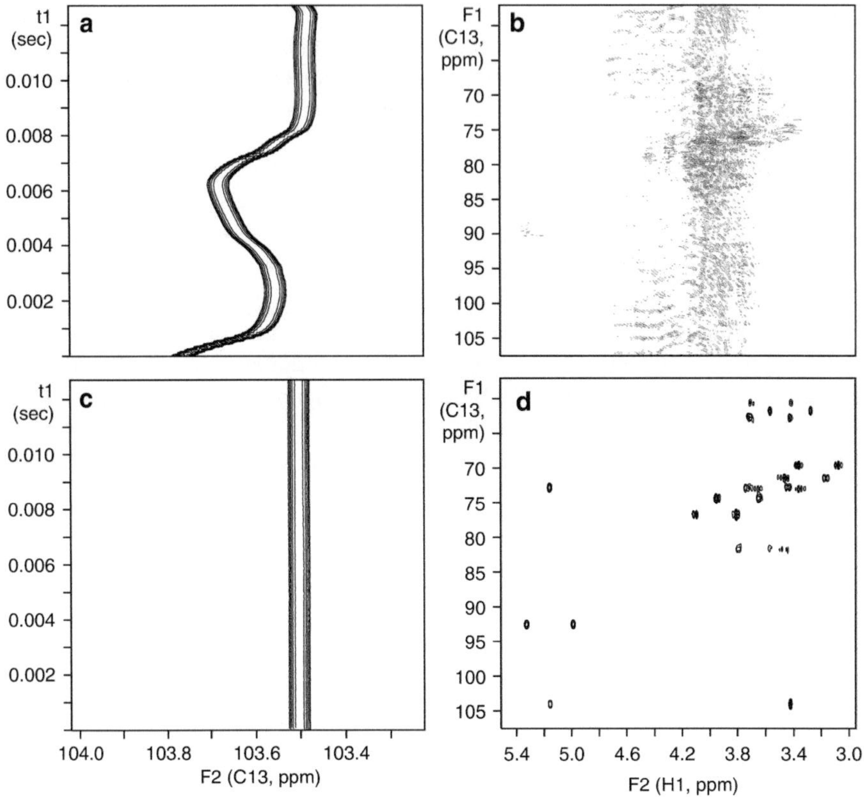

Fig. 14 The effect of internal frequency correction routine, i-Lock that is used in PANACEA experiments. The sample is sucrose in D2O. (**a**) t_1 frequency variations for one of the C-13 peaks recorded in the absence of the conventional field/frequency lock and any temperature regulation. (**b**) The corresponding two-dimensional HMBC spectrum. (**c**) The same trace as in (**a**) but after the frequency correction routine was applied. (**d**) The corresponding two-dimensional HMBC spectrum. Reproduced from Ref. 36. © Elsevier, 2010

6 Protein NMR Experiments with Simultaneous H-1 and C-13 Detection

The natural abundance of isotopes usually provides a good level of separation between various spin-systems, which is convenient when the multi-receiver experiments are designed and implemented. However, the protein NMR spectra are typically recorded with samples that are globally labelled in C-13 and N-15. A full or partial sample H-2 labelling is also frequently practised. Therefore, the multi-receiver experiments for globally labelled protein samples need to be designed by keeping this principal difference in mind. One approach to design such experiments is to start the direct detection with the least sensitive nuclei. In protein NMR this would suggest detecting the nuclear magnetization in a sequence N-15 then C13 and finally H-1.

While the vast majority of protein NMR experiments are based on proton detection, in recent years there has been a considerable interest in C-13 detection, particularly in paramagnetic proteins [38]. As a result, the cryogenic probes with optimized C-13 sensitivity and cold preamplifiers are becoming an industry standard. Hence, multi-receiver experiments that detect both C-13 and H-1 magnetization are an obvious choice. Such experiments can conveniently be designed by modifying the conventional pulse sequences. One such example based on the three-dimensional (HA)CA(CO)NNH pulse sequence [39] is discussed below.

The H-1 and C-13 detected 2D/3D (HA)CA{CO}N{NH}experiment (the curly brackets are used to indicate the nuclei that are observed directly) is shown in Fig. 15. Following the polarization transfer from protons to Cα a t_1 evolution period encodes the Cα frequencies and this magnetization is then transferred to CO for detection. In order to suppress the one-bond Cα-CO spin–spin couplings (ca. 53 Hz) the IPAP (In-Phase Anti-Phase) scheme is implemented (see panels A and B in Fig. 15). Improved resolution and simplified spectra are obtained at the expense of an additional post-processing step and a square root of two loss in sensitivity. The C-13 detected part of the experiment – the (HA)CACO sub-sequence – is shown in Fig. 15 in black and ends with acquisition of the C-13 FID. The optimum duration of the data acquisition period, t_{2C} is ca. 1.256 T_2. It can be derived from the Bloch equations and is independent of the S/N ratio (see Fig. 16) [39].

The remaining CO magnetization that has decayed too far for direct observation ("afterglow") is then refocused and transferred to protons for detection at a higher

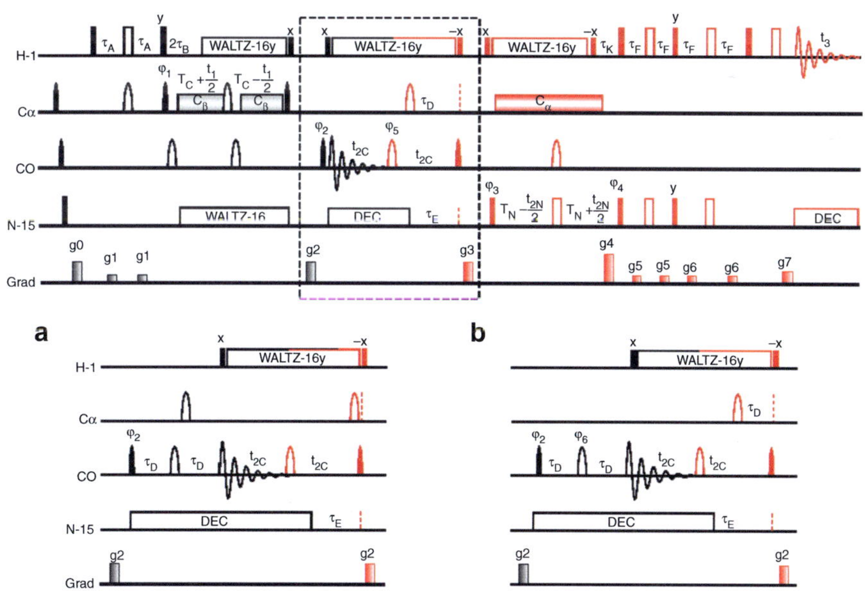

Fig. 15 The 2D/3D (HA)CA{CO}N{NH} pulse sequence. *Panels* (A) and (B) show the details of the IPAP scheme. Reproduced from Ref. 39. © The American Chemical Society, 2010

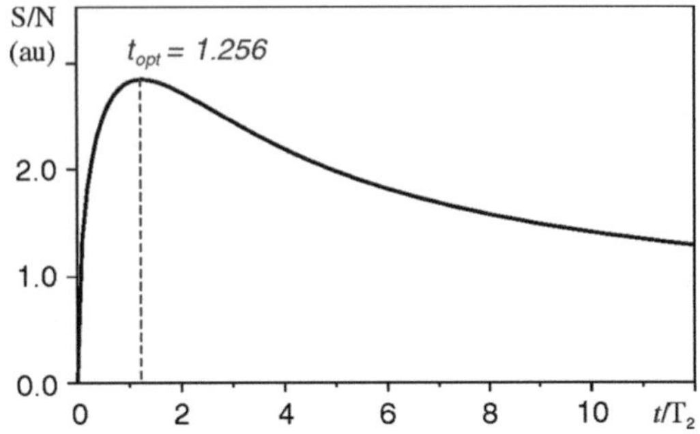

Fig. 16 The S/N ratio (in arbitrary units) as a function of the acquisition time, t and the transverse relaxation time, T_2 [39]. The optimum acquisition time, t_{opt} occurs at $t = 1.256\ T_2$. Reproduced from Ref. 39. © The American Chemical Society, 2010

sensitivity. The proton detected three-dimensional (HA)CA(CO)NNH part of the sequence that feeds on the C-13 afterglow is shown in red (see Fig. 15). First the magnetization is transferred to N-15 nuclei for further frequency encoding during the t_{2N} period. Then the N-15 magnetization is transferred to protons for detection via a sensitivity enhanced transfer step. The N-15 coherence gradients G_1 and G_2 are used to suppress unwanted coherences. The IPAP step is conveniently combined with the States-TPPI step of the N-15 hyper-complex data encoding, thus reducing the minimal duration of the experiment by a factor of two. Note that all the planes from the two-dimensional (HA)CACO experiment recorded in parallel with the three-dimensional (HA)CA(CO)NNH experiment are co-added to improve the sensitivity of the C-13 detected experiment.

Figure 17 shows the H-1 and C-13 detected spectra of a protein Nuclease A inhibitor (NuiA, 143 amino acids) recorded simultaneously in ca. 3 h. Low temperature studies of this protein (8.8 ns correlation time at 2 °C) demonstrated that moderately sized proteins can be studied using this particular pulse sequence. For proteins with sparse spectra (such as GB1, 54 amino acids) the spectra can be recorded in a considerably shorter time (ca. 15 min) using the projection reconstruction technique [4, 40].

7 Fast Techniques with Multiple Receivers

The experiments that employ multiple receivers can easily be combined with many other techniques that are designed to reduce the experiment time. The latest generation cryogenic probes provide an order of magnitude better sensitivity than the conventional room temperature probes. This implies a reduction factor of two orders of magnitude in measurement time! Of course this is only the case when the sensitivity is insufficient to record the experiment with only a few scans and significant signal averaging is required. In many cases the experiments are said to be sampling limited rather than sensitivity limited. In other words, the experiment time is limited by the data sampling schedules rather than by sensitivity, i.e. the number of scans that is required to achieve an acceptable S/N ratio. In such situations use of

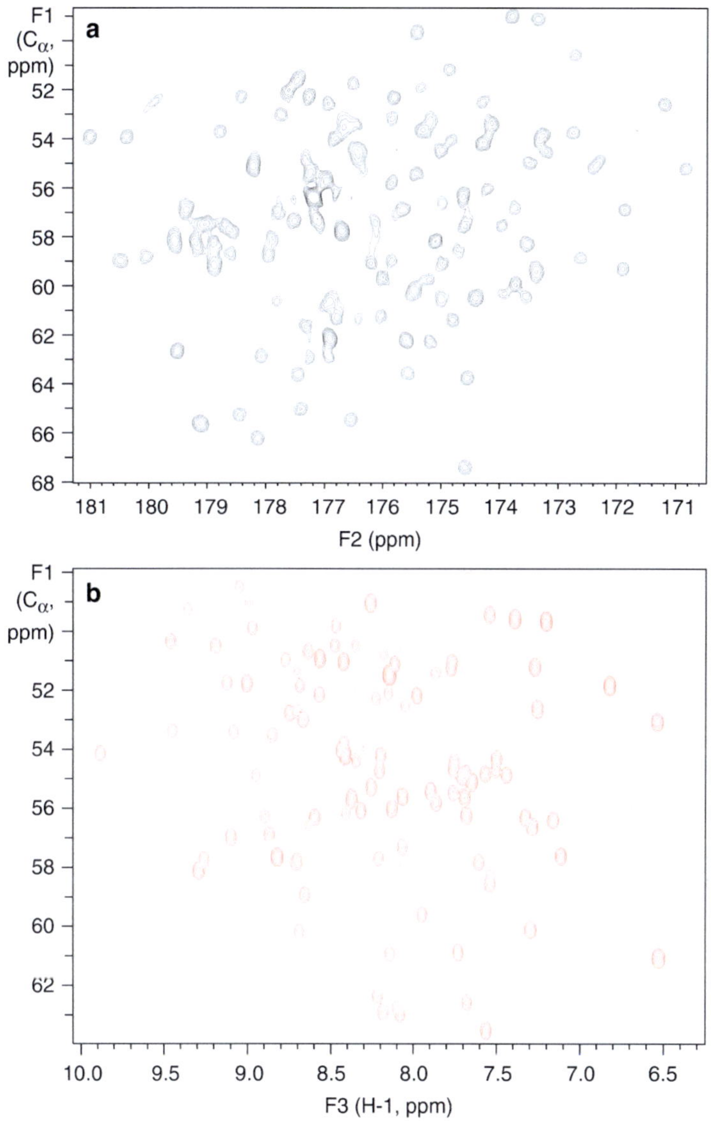

Fig. 17 Two-dimensional (HA)CACO spectrum and the F_1F_3 projection of three-dimensional (HA)CA (CO)NNH spectrum recorded simultaneously using the pulse sequence of Fig. 15. The sample is 1 mM Nuclease A inhibitor (143 amino acids). The spectra were recorded on a Varian 600 MHz NMR system equipped with two receivers and a cryogenic HCN probe optimized for C-13 detection. The experiment time was 3 h and 10 min. Reproduced from Ref. 39. © The American Chemical Society, 2010

sophisticated sampling techniques can reduce the experiment times significantly. While there are a great many fast techniques that have been developed and discussed in the literature in recent years [1–4] we shall limit our discussion to the techniques that have actually been used with multi-receiver experiments.

7.1 Hadamard NMR

Hadamard NMR spectroscopy [3] is based on the Hadamard Transform rather than the more conventional Fourier Transform. The NMR frequencies are encoded using the Hadamard matrices and multiply-selective frequency encoding pulses. Only the frequencies of interest are used at the encoding stage. Thus vast spectral regions that contain no interesting information can be excluded and the correlation spectra can be recorded much faster. Some prior knowledge about the peak positions in the indirectly detected dimensions is required for these experiments.

An example of the Hadamard encoded PANSY–TOCSY pulse sequence is shown in Fig. 18. The conventional free evolution is replaced with a Hadamard encoding pulse as can be appreciated by comparing Figs. 17 and 5. The PANSY–TOCSY spectra of inosine recorded with Hadamard encoding in just 22 s are shown in Fig. 19.

The bottleneck of the PANACEA experiment is the low sensitivity INADEQUATE section of the pulse sequence. However, given the recent advances in

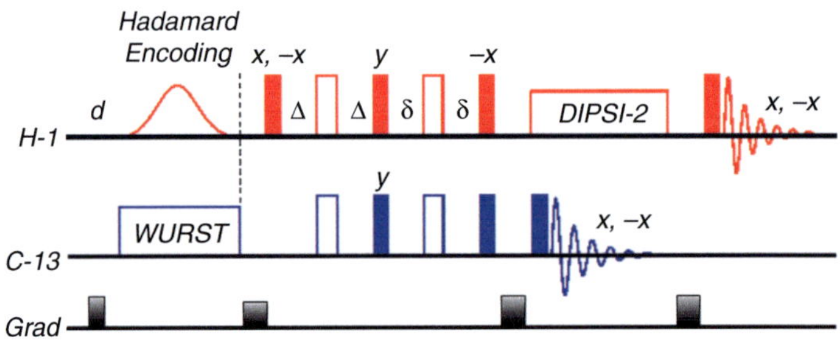

Fig. 18 PANSY–TOCSY pulse sequence with Hadamard encoding (to be compared with the conventional pulse sequence shown in Fig. 5)

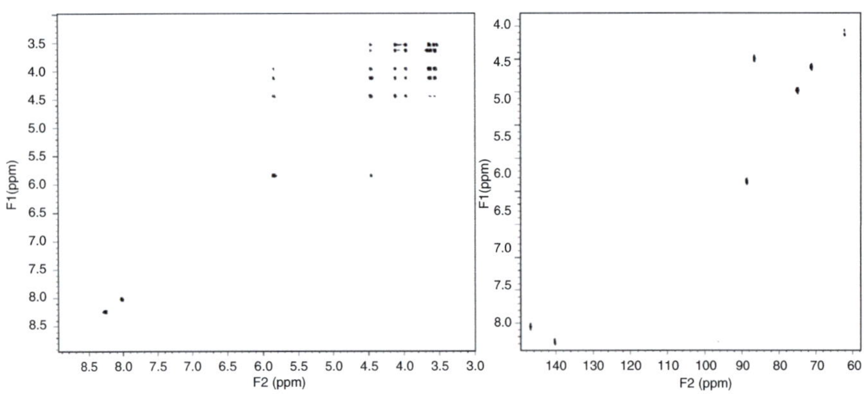

Fig. 19 Hadamard PANSY–TOCSY spectra of inosine recorded in 22 s using Hadamard encoding. A Varian 600 MHz system equipped with two receivers and a cryogenic HCN probe that is optimized for C-13 detection were used. The pulse sequence is shown in Fig. 17. Reproduced from Ref. 27. © The American Chemical Society, 2006

cryogenic probe design and the fact that the PANACEA experiment can be recorded in pure liquids, the fast techniques also apply to this rather insensitive experiment. For instance, the PANACEA spectra of menthol can also be recorded in as little as 56 s [36].

7.2 Optimized Aliasing

Computer optimized aliasing [40–42] has increased the efficiency of coherent undersampling in NMR. This technique also requires prior knowledge about the peak positions in the indirectly detected dimensions. Note that computer optimized aliasing is easily applicable to the HSQC and HMBC parts of the PANACEA experiment. However, the double quantum frequencies that would be required for optimization of the INADEQUATE experiment are generally unavailable beforehand. This is hardly a problem because the overlap in the F_1 (DQ) domain of the INADEQUATE spectra is rather unlikely and multiple aliasing is often used in practise to improve the resolution in the two-dimensional INADEQUATE spectra. Even if the signal overlap occurs, the HMBC spectra of the PANACEA experiment will usually come to the rescue and allow an unambiguous assignment of resonances. The PANACEA spectra of a cholesterol sample recorded in just 23 min are shown in Fig. 20.

As mentioned above, a rather high concentration of samples is required for fast PANACEA experiments. For instance, the PANACEA spectra of pure silicon oil have been recorded in as little as 8 min using eightfold aliasing [30, 36]. This is a nice example of using the PANACEA experiment for mixture analysis. Silicone oil

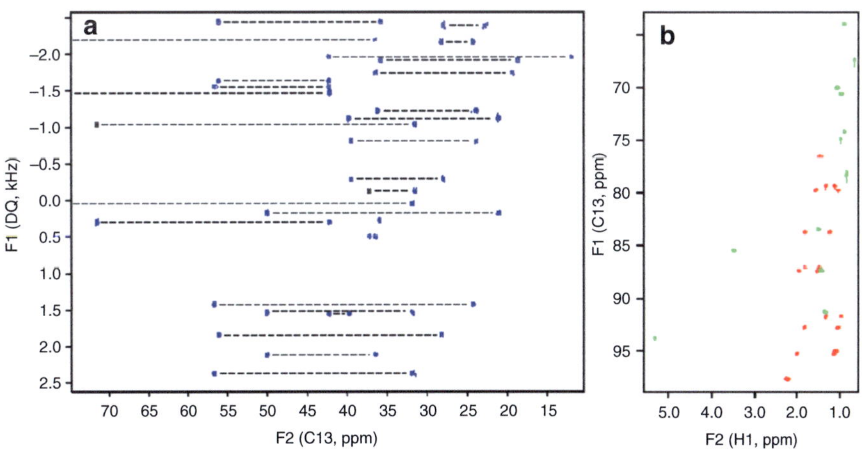

Fig. 20 Selected PANACEA spectra of a cholesterol sample (ca. 30% in $CDCl_3$) recorded on a Varian 600 MHz NMR system equipped with two receivers and a cryogenic HCN probe optimized for C-13 detection. The experiment time was 23 min. A fourfold undersampling (aliasing) was used with optimized offset and spectral window in the F_1 (SQ) dimension. (**a**) Inadequate spectrum; (**b**) multiplicity edited HSQC spectrum (the negative peaks are shown in *red*). Reproduced from Ref. 36. ©Elsevier, 2010

usually consists of a mixture of silicone chains of different length. The considerably higher natural abundance of the Si-29 nuclei (4.74%) and favourable polarization transfer conditions from abundant methyl groups provide for a better sensitivity. On the other hand, small two-bond J(Si–Si) couplings of approximately 2 Hz introduce long inter-pulse delays and contribute to higher relaxation losses. The relatively long relaxation times of Si-29 nuclei help to minimize such losses and overall the PANACEA experiment can be recorded in just a few minutes. In this case the one-dimensional Si-29 spectrum also serves as i-Lock that replaces the conventional deuterium frequency/field lock. The heteronuclear Si–C couplings can easily be measured from these spectra and provide further structural information.

7.3 *Projection Reconstruction Experiments*

The projection reconstruction (PR) method originates from imaging techniques [43, 44] and is based on the projection theorem [45, 46]. The technique has recently attracted considerable interest in multi-dimensional NMR spectroscopy [47–49], especially in the field of protein NMR [2–4, 17–19]. It is a particularly efficient time saving technique that works amazingly well with sparse spectra and medium size proteins. The technique has been applied in combination with the 2D/3D (HA)CA {CO}N{NH} experiment to a small protein, GB1 [36]. The three-dimensional experiment was recorded in just 15 min. The three-dimensional spectrum was reconstructed from three planes – two orthogonal planes and a tilted plane recorded at an optimum projection angle of 69.5°. Figure 21 shows the comparison of the

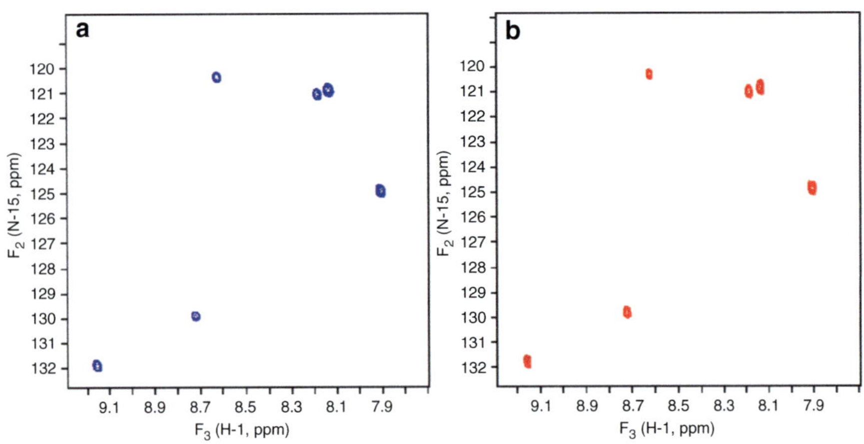

Fig. 21 Comparison between the F_2F_3 planes at $F_1 = 51.93$ ppm of (**a**) the conventional 3D (HA) CA(CO)NNH spectrum of GB1 recorded in 190 min and (**b**) the reconstructed 3D spectrum recorded using the PR technique in just 15 min. In both cases the ^{13}C detected (HA)CACO experiment was simultaneously acquired. Reproduced from Ref. 39. ©The American Chemical Society, 2010

F_2F_3 plane extracted at $F_1 = 51.93$ ppm from the reconstructed spectrum and a spectrum recorded in 3 h and 10 min using the conventional sampling.

It should be noted that both the two-dimensional (HA)CACO and three-dimensional (HA)CA(CO)NNH sections of the experiment share the Cα evolution period, t_1. In the PR experiment the tilted plane is recorded by jointly incrementing the (t_1, t_2) evolution times according to $t_1 = \cos\delta/\mathrm{sw}_{\mathrm{tilt}}$ and $t_2 = \sin\delta/\mathrm{sw}_{\mathrm{tilt}}$, where δ is the tilt (projection) angle and the $\mathrm{sw}_{\mathrm{tilt}}$ is the spectral width in the tilted plane. Since the C-13 FID that is recorded during t_{2C} is independent of the subsequent events in the pulse sequence, the F_1 axis in the (HA)CACO spectrum needs be scaled by 1/ $\cos\delta$. This should be taken into account if several such planes need to be combined.

8 Data Processing in Multi-Receiver Experiments

There are certain restrictions that, at the time of writing this chapter, limit the flexibility of multi-receiver data acquisition and consequently require a few additional processing steps that need to be implemented before the conventional data processing can be applied. First of all the multi-receiver data are acquired in interleaved fashion, meaning that the data need to be separated prior to further processing. To some extent this is a question of convenience because software solutions for processing raw multi-receiver data do exist. Still, porting the data into other processing tools may require some pre-processing (see Fig. 22). In applications where i-Lock is used such pre-processing would start with frequency correction followed by data separation.

Multi-receiver data are often recorded for nuclei that have vastly different spectral widths, for instance H-1/C-13 or H-1/F-19. For convenience the heavily

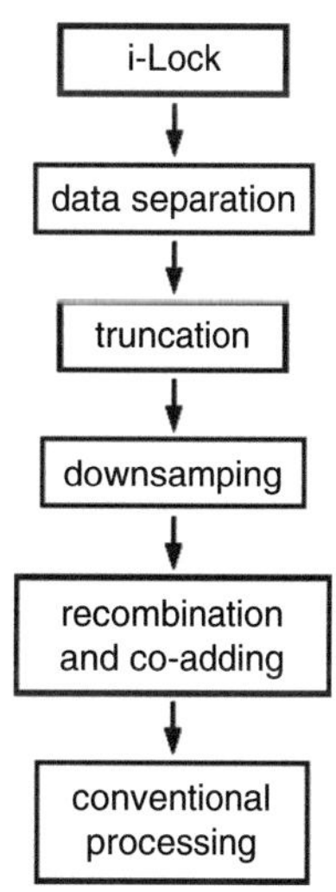

Fig. 22 The processing flow chart for handling data recorded using multiple receivers

oversampled H-1 spectra should be down-sampled to a more acceptable spectral width. This not only reduces the data storage size but also makes the subsequent processing and data visualization faster and more convenient. The same is true for situations when the T_2 relaxation times of various nuclear species are significantly different. Again, truncating the data sets of faster relaxing nuclei in the time domain is useful and serves similar purposes.

Extracting the spin–spin coupling information may require use of high levels of digitization, which may involve manipulating huge data matrices. In such cases partial data processing proves to be rather convenient. The particular spectral window is extracted from the full data matrix and properly digitized, processed and displayed as required. In this way highly accurate values of long range couplings have been extracted from the 3D PANACEA data sets [35]. For instance, the long-range couplings involving the OH-group in Me-salicylate have been measured with the accuracy of ± 0.05 Hz using the J-doubling technique [50] (see Fig. 23). The information about the orientation of the OH group in this molecule extracted from such measurements suggests a preferential orientation and hydrogen bonding.

When the dimensionality of the spectra in a particular experiment differs, the pre-processing involves adding multiple spectra of lower dimensionality (usually traces or planes). For instance, all one-dimensional spectra recorded using the

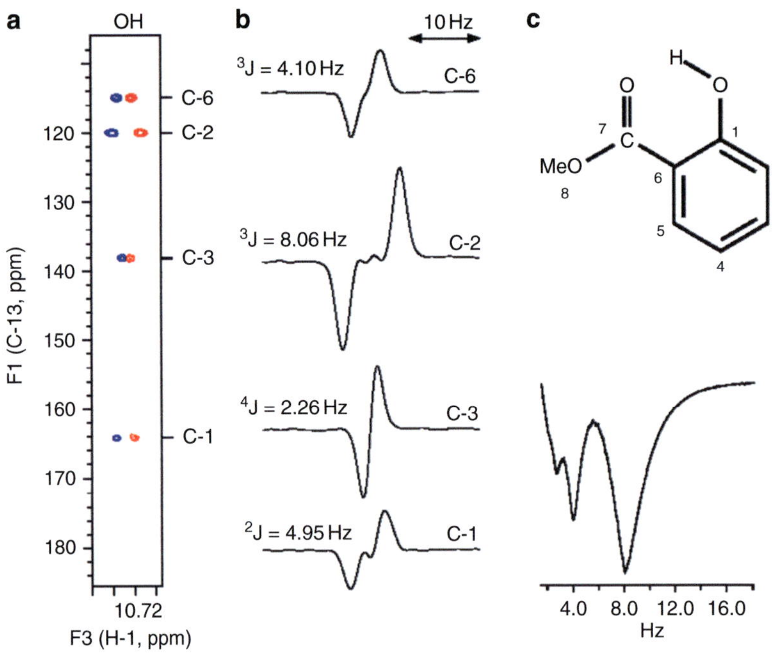

Fig. 23 The J(C,OH) spin-couplings extracted from the PANACEA spectra of Me-salicylate. (**a**) The OH region of the J-HMBC spectrum; (**b**) the corresponding traces; (**c**) $^3J(C_2,OH)$ measurement by cascaded J-doubling. The deep minimum indicates a splitting of 8.06 ± 0.05 Hz. The two weaker minima fall at sub-harmonics of this frequency [36]

PANACEA pulse sequence can be added together, providing a spectrum with high S/N ratio that allows measurement of spin–spin coupling constants and thus extracting additional structural information. Similarly, the multiple two-dimensional Cα-CO correlation spectra that are recorded during the three-dimensional (HA)CA (CO)NNH experiment are co-added to improve the S/N ratio. In fact, this happens "automatically" as a result of three-dimensional Fourier transform.

9 Conclusions

The recent introduction of multiple receiver technology in NMR spectroscopy opens a new avenue in the field of NMR experiment design. The multi-receiver experiments introduced so far have demonstrated the utility of this technology. Such experiments have been applied to H-1, C-13, N-15, F-19, Si-29 and P-31 NMR and have proved to provide considerably more information in a single measurement as compared to conventional NMR spectroscopy that uses single receiver systems. A new type of experiment (PANACEA) that allows unambiguous structure determination of small organic molecules has been introduced. Such experiments work in pure liquids, i.e. without the conventional field/frequency lock system. On multi-receiver systems essentially any nucleus can be used for the internal frequency stabilization (i-Lock). More recently new experiments with simultaneous H-1 and C-13 detection have been introduced in bio-molecular NMR. The experiments described here can easily be combined with many fast acquisition techniques. This provides significant time savings in multi-dimensional multi-receiver experiments.

References

1. Gal M, Frydman L (2010) Multidimensional NMR methods for the solution state. In: Morris GA, Emsley JW (eds) Encyclopedia of magnetic resonance, Chap. 3. Wiley, Chichester, UK
2. Kupče Ē, Freeman R (2003) J Biomol NMR 27:101–113
3. Kupče Ē, Nishida T, Freeman R (2003) Progr NMR Spectrosc 42:95–122
4. Kupče Ē, Freeman R (2006) In: Arrondo JLR, Alonso A (eds) Advanced techniques in biophysics, Chap. 6. Springer, Berlin, pp 131–147
5. Ding K, Gronenborn A (2002) J Magn Reson 156:262–268
6. Hiller S, Fiorito F, Wüthrich K, Wider G (2005) Proc Natl Acad Sci USA 102:10876–10881
7. Fiorito F, Hiller S, Wider G, Wüthrich K (2006) J Biomol NMR 35:27–37
8. Brüschweiler R, Zhang F (2004) J Chem Phys 120:5253–5260
9. Zhang F, Brüschweiler R (2004) J Am Chem Soc 126:13180–13181
10. Frydman L, Scherf T, Lupulescu A (2002) Proc Natl Acad Sci USA 99:15858
11. Frydman L, Scherf T, Lupulescu A (2003) J Am Chem Soc 125:9204
12. Barna JCJ, Laue ED, Mayger MR, Skilling J, Worrall SJP (1987) J Magn Reson 73:69–77
13. Chen J, Mandelshtam VA, Shaka AJ (2000) J Magn Reson 146:363–368
14. Schmieder P, Stern AS, Wagner G, Hoch JC (1993) J Biomol NMR 3:569

15. Kazimierczuk K, Zawadzka A, Kozminski W, Zhukov I (2006) J Biomol NMR 36:157–168
16. Kazimierczuk K, Kozminski W, Zhukov I (2006) J Magn Reson 179:323–328
17. Kupče Ē, Freeman R (2003) J Am Chem Soc 125:13958–13959
18. Kupče Ē, Freeman R (2004) J Am Chem Soc 126:6429–6440
19. Yoon JW, Godsill S, Kupče Ē, Freeman R (2006) Magn Reson Chem 44:197–209
20. Schanda P, Brutscher B (2005) J Am Chem Soc 127:8014
21. Schanda P, Kupče Ē, Brutscher B (2005) J Biomol NMR 33:199
22. van de Ven FJM (1995) Multidimensional NMR in liquids. VCH, New York
23. Cavanagh J, Fairbrother WJ, Palmer AG III, Skelton NJ (1996) Protein NMR spectroscopy. Academic, San Diego
24. Moore GJ, Hrovat MI, Gonzalez RG (1991) Magn Reson Med 19:105–112
25. Hou T, MacNamara E, MacNaughton M, Raftery D (1999) Anal Chim Acta 400:297–305
26. Blaimer M, Breuer F, Mueller M, Heidemann RM, Griswold MA, Jakob PM (2004) Top Magn Reson Imaging 15:223–236
27. Kupče Ē, Freeman R, John BK (2006) J Am Chem Soc 128:9606–9607
28. Ernst RR, Bodenhausen G, Wokaun A (1997) Principles of nuclear magnetic resonance in one and two dimensions. Clarendon, Oxford
29. Robinson JN, Coy A, Dykstra R, Eccles CD, Hunter MW, Callaghan PT (2006) J Magn Reson 182:343–347
30. Kupče Ē, Wrackmeyer B (2010) Appl Organomet Chem (Spl Issue: In Memoriam Professor Edmunds Lukevics) 24:837–841
31. Kupče Ē, Cheatham S, Freeman R (2007) Magn Reson Chem 45:378–380
32. Bax A, Freeman R, Kempsell SP (1980) J Am Chem Soc 102:4849–4851
33. Bax A, Freeman R, Kempsell SP (1980) J Magn Reson 41:349–353
34. Kupče Ē, Freeman R (2008) J Am Chem Soc 130:10788–10792
35. Kupče Ē, Freeman SR (2010) Magn Reson Chem 48:333–336
36. Kupče Ē, Freeman R (2010) J Magn Reson 206:147–153
37. Iijima T, Takegoshi K (2008) J Magn Reson 191:128–134
38. Bermel W, Bertini I, Felli IC, Piccioli M, Pierattelli R (2006) Prog NMR Spectrosc 48:25
39. Kupče Ē, Kay LE, Freeman R (2010) J Am Chem Soc 132:18008–18011
40. Jeannerat D (2003) Magn Reson Chem 41:3–17
41. Jeannerat D (2007) J Magn Reson 186:112–122
42. Lescop E, Schanda P, Rasia R, Brutscher B (2007) J Am Chem Soc 129:2756–2757
43. Hounsfield GN (1973) Br J Radiol 46:1016
44. Zhi-Pei Liang, Lauterbur PC (2000) Principles of magnetic resonance imaging, Chap. 6. IEEE, New York, pp 187–216
45. Nagayama K, Bachmann P, Wuthrich K, Ernst RR (1978) J Magn Reson 31:133
46. Bodenhausen G, Ernst RR (1982) J Am Chem Soc 104:1304–1309
47. Kupče Ē, Freeman R (2004) Concepts Magn Reson 22A:4–11
48. Kupče Ē, Freeman R (2004) Concepts Magn Reson 23A:63–75
49. Kupče Ē, Freeman R (2004) Spectroscopy 19(10):16–20
50. McIntyre L, Freeman R (1992) J Magn Reson 96:425

Top Curr Chem (2013) 335: 97–120
DOI: 10.1007/128_2011_228

Published online: 17 September 2011

TROSY NMR Spectroscopy of Large Soluble Proteins

Yingqi Xu and Stephen Matthews

Abstract Solution nuclear magnetic resonance spectroscopy is usually only used to study proteins with molecular weight not exceeding about 50 kDa. This size limit has been lifted significantly in recent years, thanks to the development of labelling methods and the application of transverse-relaxation optimized spectroscopy (TROSY). In particular, methyl-specific labelling and methyl-TROSY have been shown to be effective for supramolecular systems as large as about 1 MDa. In this chapter we review the available methods for labelling different kinds of methyl groups and the assignment strategies in very large protein systems. Several proteins are selected as examples to show how NMR helps to reveal the details of structure, interaction and dynamics of these proteins.

Keywords Large proteins · Methyl labelling · Mutagenesis · NOESY · TROSY

Contents

Y. Xu and S. Matthews (✉)
Division of Molecular Biosciences, Imperial College London, South Kensington, London SW7 2AZ, UK
e-mail: s.j.matthews@imperial.ac.uk

Abbreviations

ATCase	Aspartate transcarbamoylase
CSA	Chemical shift anisotropy
HMQC	Heteronuclear multiple-quantum correlation
HSQC	Heteronuclear single-quantum correlation
NMR	Nuclear magnetic resonance
NOE	Nuclear Overhauser effect
PCS	Pseudocontact shift
PRE	Paramagnetic relaxation enhancement
RDC	Residual dipolar coupling
TROSY	Transverse relaxation optimized spectroscopy

1 Introduction

Compared with X-ray crystallography, solution NMR spectroscopy is usually limited to the study of relatively small proteins. For larger proteins, a 'divide and conquer' strategy is often used: domains (independent structural and functional modules) identified from the proteins are studied separately and the information is then integrated to gain insight into the whole molecule. However, it is arguable that such a picture does not reflects the real scenario, especially when multiple domains need to cooperate with each other to carry out the function of interest. The details of interactions between these domains can only be obtained by studying the large complex. In addition, many proteins and protein complexes are difficult to break apart. Fortunately, within the past decade or so we have witnessed great progress in NMR methodology as well as better NMR equipment so that the size of molecules amendable to NMR studies has increased from ~25 kDa [1, 2] to nearly 1 MDa [3]. NMR spectrometers with higher magnetic fields and cryogenic probes have helped to generate spectra for large proteins, with better resolution and sensitivity. Deuterated samples, in which most protons are replaced with deuterons, can reduce the relaxation rates significantly [4]. Most importantly, as we will review here, methods have been established to record spectra of deuterated samples with higher sensitivity and resolution by selecting and preserving the slowly relaxing components of NMR signals, so that much larger proteins than before can now be studied.

Transverse-relaxation optimized spectroscopy (TROSY) [5], as its name suggested, seeks to optimize (minimize) the relaxation of transverse signals. Transverse signals are the observable (directly or indirectly) signals in modern NMR, and their relaxation leads to signal decay, or widening on the Fourier-transformed spectrum. The transverse relaxation rate increases roughly in linear relationship with the molecular weight of a protein. The accelerated relaxation rate is the main reason why NMR is so difficult for large proteins. Different mechanisms contribute to relaxation; in a few cases these contributions cancel each other out, resulting in slower relaxation, which can be selectively retained and observed by TROSY-type experiments.

The first TROSY was demonstrated by Pervushin et al. in 1997 [5], producing similar information as a classical heteronuclear single-quantum correlation (HSQC) experiments [6] for amide groups, but with much higher sensitivity and resolution for large deuterated proteins. The two dominant mechanisms contributing to relaxation of an amide $^{15}N/^{1}H$ are: (1) dipolar coupling between ^{15}N and ^{1}H, and (2) chemical shift anisotropy (CSA) of $^{15}N/^{1}H$. A $^{15}N/^{1}H$ nucleus has two possible spin states in a magnetic field, with one pointing in the same direction as the magnetic field and the other pointing in the opposite direction. The direction of dipole–dipole interaction depends on the spin states of both nuclei involved, while the direction of CSA depends only on the spin state of the nucleus itself. It happens that these two contributions are nearly co-linear; therefore, in one half of the amides, they add up, while in another half, they cancel out. The difference in relaxation rates can be seen from the different widths of the four components of an amide peak (Fig. 1b). Only one component has slower relaxation (narrower width) for both ^{15}N and ^{1}H, which is selectively kept in TROSY. A conventional HSQC experiment includes decoupling pulses, which mix up the four components and lead to all the signals decaying efficiently in large proteins. Because the strength of CSA relates to field strength, it is expected that at ~1 GHz the TROSY effect is most significant, where the two mechanisms have similar strength. After the initial report of 2D TROSY, the technique has been quickly incorporated into all triple resonance experiments [7] for backbone assignments, leading to NMR studies of molecular systems well above 100 kDa [8–10]. In addition to amides, the TROSY effect was also found for some aromatic side chains [11].

More recently, Tugarinov et al. showed that similar cross-correlations exist in side chain methyl relaxations [12]. For an isolated methyl group, the main contributions to relaxation are from dipole–dipole interactions with intramethyl protons. Considering the magnetization involving ^{13}C and one ^{1}H, the dipolar interactions with the other two ^{1}H, which lead to relaxation, will cancel out when those two ^{1}H have opposite spin states. This is the case for half of the methyl group population; therefore, half of the signal can be preserved in methyl-TROSY, which is effectively the traditional ^{1}H-^{13}C HMQC [6], whose superiority to HSQC in large proteins is easily seen from Fig. 2. In contrast to amide TROSY, methyl TROSY does not rely on a high magnetic field. Because the methyl groups are located at the far end of amino acid side chains and rotate rapidly, they usually have more favourable relaxation characteristics than amides so that methyl-TROSY has the potential to be used on systems where amide-TROSY fails. In fact, high quality methyl-TROSY spectra have been recorded on supralarge systems with molecular weight up to ~1 MDa [3, 13], more than an order of magnitude larger than

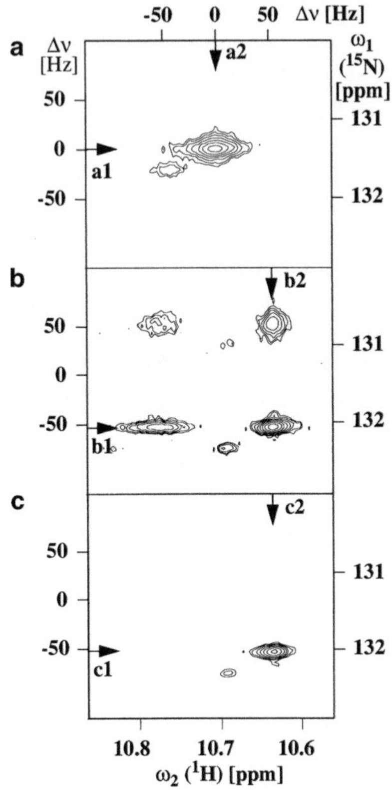

Fig. 1 Contour plots of ^{15}N, ^{1}H correlation spectra for an example ^{15}N-^{1}H spin system [5]. (**a**) Conventional broad-band decoupled [^{15}N,^{1}H]COSY spectrum. (**b**) Conventional [^{15}N,^{1}H]COSY spectrum recorded without decoupling. (**c**) TROSY-type ^{15}N,^{1}H correlation spectrum. Reprinted from [5]

conventional NMR applications. In this review, we discuss recent progress in this direction. Readers interested in detailed principles behind the TROSY experiments are referred to the original papers [5, 12] as well as some earlier reviews [9, 14, 15]. We will focus on the general routines of successful applications to supramolecular systems, the insights gained through these studies, and possible future developments.

2 Preparation of Methyl-Labelled Samples

The first step of any NMR study is to make suitable samples. For large proteins, this is usually not trivial. NMR experiments need highly concentrated samples (~1 mM), which means that for bigger proteins more protein molecules have to be dissolved into a small volume. Aggregation has to be carefully avoided while solubility and stability are optimized by screening for the best buffer condition and temperature. The cost is always much higher than preparing small protein samples because deuteration is needed to remove of all the protons (except for the methyl

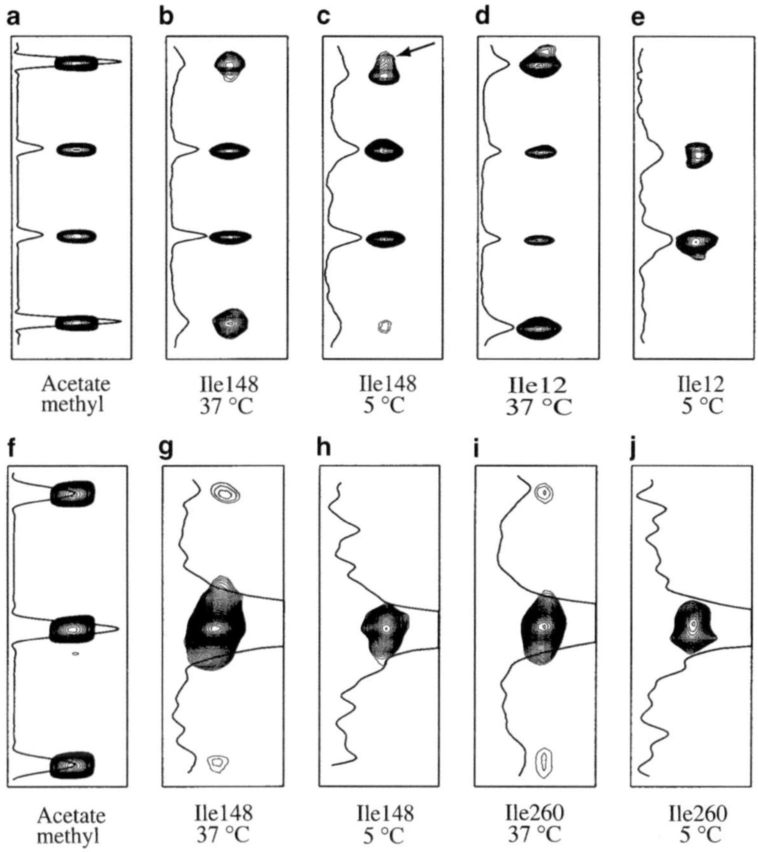

Fig. 2 ^{13}C-multiplet structures of selected Ile δ1 peaks of MSG and the methyl group of sodium acetate from ^{1}H-coupled HSQC (**a**–**e**) and HMQC (**f**–**j**) spectra. 1D traces taken at each peak position are shown. 'Quartets' of (**a**) the methyl of sodium acetate (approximate intensity ratio in the quartet 4:1:1:4); (**b**) Ile148 at 37 °C; (**c**) Ile148 at 5 °C (the outer fast-relaxing lines are still visible but are significantly reduced in intensity compared to 37 °C; the *arrow* indicates the position of the outer peak, partially overlapped with a peak from another HSQC multiplet); (**d**) Ile12 at 37 °C; and (**e**) Ile12 at 5 °C (only the two slowly relaxing inner lines are detected). 'Triplets' of (**f**) the methyl of sodium acetate (approximate intensity ratios of 1:1.8:1); (**g**) Ile148 at 37 °C; (**h**) Ile148 at 5 °C (only the central line, L2′, is visible); (**i**) Ile260 at 37 °C; and (**j**) Ile260 at 5 °C (only L2′ is visible). Since the intensities of the outer lines of the HMQC mutiplets (L1′, L3′) are extremely small relative to the inner line, L2′ (<3.5% at 37 °C and only observed for Ile148 and Ile260), the HMQC spectra are drawn at ~tenfold lower contour levels than their HSQC counterparts. Reprinted from [12] with permission, copyright 2003 ACS

protons of interest) to achieve the maximal effect of methyl TROSY. Deuteration is realized by growing cells in 100% D_2O, which leads to slower growth rate and often lower protein yield. Several protocols have been reported to optimize the yield of deuterated proteins [16, 17].

Over the years, different methods have been developed for selective methyl protonation. An early trial used protonated pyruvate as the sole carbon source [18] and produced samples highly protonated at the methyl sites of Ala, Val, Leu and Ile (γ2 only), while most other sites were highly deuterated. Recently, it has been reported that using protonated glucose as the sole carbon source [19] resulted in ~50% of all the methyl groups labelled as CHD_2, which could be selectively detected with a modified constant-time ^{1}H-^{13}C HSQC. These methods can be useful for studies of proteins below ~40 kDa, whereas for larger proteins, labelling methods with very little scrambling or isotopomers and more complete deuteration background are preferred. For such labelling protocols, deuterated glucose is used together with 100% D_2O to achieve very high deuteration; suitable precursor(s) for methyl protonation are added into the media ~1 h before induction.

2.1 Ile, Leu and Val Labelling

Many applications used selective methyl protonation of Ile (δ1 only), Leu and Val (ILV), which was developed in Kay's laboratory [16, 20]. By supplementing the minimal media with 50 mg/L of (3,3-$^{2}H_2$) 2-ketobutyrate [21], the δ1 methyl groups of Ile were selectively protonated; further supplementing the media with 50 mg/L Val resulted in fully protonated methyls in Val and Leu residues. Inclusion of ~100 mg/L (3-$^{2}H_2$) 2-ketoisovalerate [22] instead of Val offered better labelling of Leu methyl groups and proved to be more cost-effective. More than 90% of methyl groups were labelled as CH_3 using these methods, offering spectra with high resolution as well as high sensitivity. Depending on the NMR experiments needed, all the carbons or only the methyl carbon(s) may be labelled with ^{13}C. For application to very high molecular weight proteins using methyl-TROSY experiments, it is better to use 2-keto-3-methyl-d_3-3-d_1-butyrate [23] so that only one methyl from Val or Leu will be labelled while the other methyl will be NMR-inactive (i.e. U-[^{2}H]-Leu/Val-[$^{13}CH_3$, $^{12}CD_3$]). Although this almost halves the detectable methyl concentration, these methyls do not suffer from relaxation caused by intraresidue methyl protons, and the experiments correlating methyl group with other side chain ^{13}C resonances (see Sect. 3.1) have better sensitivity because the magnetization transfer is not diverted.

Because the spectra tend to be overcrowded for high molecular weight proteins, it is desirable to label only the *pro-S* methyl groups of Leu or Val, using a newly introduced method to synthesize (*S*)-2-acetolactate, whose 2-[$^{13}CH_3$] is stereospecifically transferred to the *pro-S* position of α-ketoisovalerate in vivo [24]. In addition to simplifying the spectra so that only 50% signals are observable, the remaining 50% of the methyl groups are fully protonated instead of being diluted and are therefore more easily detected (see Fig. 3). This is crucial for proteins that cannot be highly concentrated and may generate nuclear Overhauser effect (NOE) signals among more distant groups.

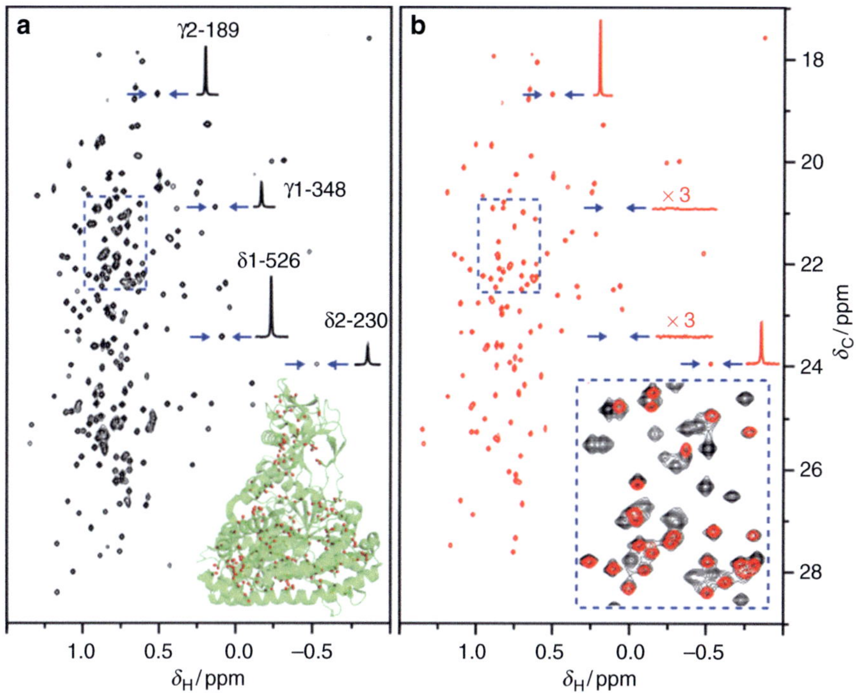

Fig. 3 Comparison of methyl-TROSY spectra of MSG samples with specifically labelled Leu and Val methyl groups: (**a**) U-[^{2}H], U-[^{12}C], Leu/Val-[$^{12}C^2H_3$,$^{13}C^1H_3$] MSG with non-stereospecific methyl labelling prepared by using 3-[$^2H_3$]methyl-3-[^{2}H]-4-[^{13}C]-α-ketoisovalerate; (**b**) U-[^{2}H], U-[^{12}C], Leu/Val-[$^{13}C^1H_3$]$^{pro\text{-}S}$ MSG with stereospecific labelling prepared by using 2-[^{13}C] methyl-4-[$^2H_3$]acetolactate. Examples of 1D traces are shown at the positions indicated by *arrows*. Correlations of Leu-δ_1 and Val-γ_1 (pro-*R* methyl groups) are not detected in (**b**), whereas signals for Leu-δ_2 and Val-γ_2 (*pro-S* methyl groups) are enhanced by a factor of 1.6 ± 0.4. The *inset* in (**a**) shows the location of Leu and Val methyl groups (*balls*) on the 3D structure of MSG. An overlay of a region of the spectra (indicated by *dashed boxes*) is presented in the *bottom right corner* of (**b**). Adapted from [24] with permission

Very recently, a method for ^{13}C,^{1}H labelling at the Ile-γ2 position was also reported, using 140 mg/L α-aceto-α-hydroxybutyrate as precursor [25]. Although some *proR* methyl positions of Leu and Val are 'accidentally' labelled, the level is relatively low (~4%) and does not affect the quality of spectra as the overlap is minimal.

The methyl groups can also be labelled as $^{13}CHD_2$ instead of $^{13}CH_3$. Because the intramethyl dipolar relaxation is eliminated de facto, this provides a simplified spin system for relaxation studies [26]; however, the sensitivity is lower than for CH_3 probes by a factor of 1.5–2 [27], indicating that CH_3 probes should be the method of choice for most applications.

2.2 Labelling of Ala and Met

In addition to Ile, Leu and Val, Ala is another abundant amino acid in proteins and is evenly distributed in the protein structure, both in the hydrophobic cores and on protein surfaces. Labelling of Ala methyl groups is not as convenient as labelling of ILV methyl groups because it always suffer from scrambling. However, by adding 50 mg/L of L-alanine-3-^{13}C,2-^{2}H into a deuterated rich media [28] instead of minimal media, Ala methyls could be labelled to comparable levels as Ile methyls with no significant scrambling (see Fig. 4). It is worth noting that the alanine precursor is not commercially available, but can be prepared from protonated alanine with tryptophan synthase to accelerate the exchange of α-^{1}H into ^{2}H. It has also been reported that, by addition of perdeuterated forms of α-ketoisovalerate, succinate and L-isoleucine together with 800 mg/L L-alanine into minimal media, undesired background labelling can be reduced to <1%, thereby providing a more cost-effective way for Ala methyl labelling [29]. Due to the proximity of Ala β-methyl groups to the backbone, a majority of them are highly ordered [30], making them especially useful for the measurement of methyl ^{1}H-^{13}C residual dipolar couplings (RDC) [31].

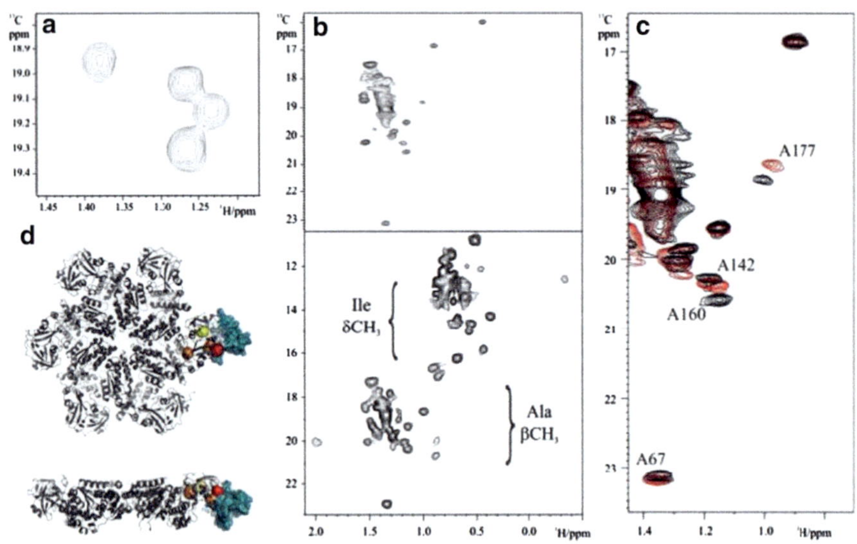

Fig. 4 Alanine labelling and ^{1}H-^{13}C 2D methyl-TROSY spectra of (**a**) ^{13}CH$_3$-Ala,U-^{2}H-labelled Ufd1 in Ufd1/Npl4 complex, and (**b**) ^{13}CH$_3$-Ala,U-^{2}H-labelled p97 ND1 (*top*) and ^{13}C$_\delta$H$_3$-Ile, ^{13}CH$_3$-Ala,U-^{2}H-labelled p97 ND1 (*bottom*). (**c**) Expanded region of the ^{1}H-^{13}C 2D methyl-TROSY spectrum of ^{13}CH$_3$-Ala,U-^{2}H-labelled p97 ND1 in the absence (*black*) and presence (*red*) of perdeuterated Npl4 UBD. (**d**) Structural model of the p97 ND1-Npl4 UBD complex. The p97 hexamer is shown as a *grey ribbon* and perturbed alanine residues are indicated by *red* (most shifted), *orange* (intermediate) and *yellow* (least) balls. The position of Npl4 UBD (*blue surface*) in complex with p97 N domain is shown for one of monomers. Reprinted from [28] with permission, copyright 2007 ACS

Table 1 Amino acids containing methyl groups and their labelling method

Amino acid	Frequency in proteins (%)	Labelling method (precursor)
Isoleucine	5.2	$^{13}CH_3$–$^{12}CD_2$–^{12}CO–$^{12}COO^-$ for δ1 $^{13}CH_3$–^{12}CO–($^{12}CD_3$–$^{12}CD_2$)$^{12}C(OD)$–$^{12}COO^-$ for γ2
Leucine	9.0	$^{13}CH_3$-^{12}CD($^{12}CD_3$)-^{12}CO-$^{12}COO^-$
Valine	6.6	–$^{12}CD_3$-^{12}CO-($^{13}CH_3$)$^{12}C(OD)$-$^{12}COO^-$ for *pro-S* methyl labelling
Methionine	2.4	$^{13}CH_3$–S–$^{12}CH_2$–$^{12}CH_2$–$^{12}CH(NH_3^+)$–$^{12}COO^-$
Alanine	8.3	$^{13}CH_3$–$^{12}CD(NH_3^+)$–$^{12}COO^-$
Threonine	5.8	?

Labelling of Met S-^{13}C-methyl was also reported using 65 mg/L [4-^{13}C] methylthio-2-oxobutanoate [32] or 250 mg/L [$^{13}CH_3$]-methionine [13] as a precursor. There are usually not as many methionine residues as the other amino acid types in a protein, making it extremely attractive when one needs to minimize the observable signals to focus on a few desired locations. Another advantage of the ε-CH_3 from methionine is that it usually has more flexibility and generates sharper NMR peaks, which is useful when the signals from other methyl groups are too broad to be observed [33].

A brief list of the labelling methods for different kinds of methyl groups can be found in Table 1. Except for Thr, which is upstream of several amino acid biosynthetic pathways, all the amino acids containing methyl groups can be effectively labelled, providing very good 'coverage' of the protein for probing its structure and dynamics. Labelling of Thr methyls should also be possible by using cell-free expression [34, 35], where the scrambling is minimized.

3 Assignment of the Methyl Resonances

When a suitable sample has been prepared, a 2D methyl-TROSY can be recorded to see whether the peaks have enough sensitivity and resolution. If the TROSY spectrum looks promising, it can be used to probe the interactions and dynamics of the protein. However, such details can only be fully explained after the peaks have been assigned to specific methyl groups in the protein sequence and mapped onto the structure.

3.1 Correlation with Assigned Backbone Resonances

Typically, for smaller proteins, the backbone (namely HN, N, Cα and C') and Cβ resonances are assigned first; the Ile δ1, Leu δ and Val γ methyl resonances are then correlated with backbone amide resonances using HN-detected experiments or with

Cα, Cβ and C’ using out-and-back methyl-detected experiments [36]. To ensure that magnetization transfers more efficiently, a COSY-based scheme using selective refocusing pulses was utilized [37]. In addition, for Leu and Val, one of the two methyl groups was labelled as $^{12}CD_3$ so that it would not interfere with magnetization transfer from the other methyl. Similar correlation experiments for assignment of Ala β and Ile γ2 methyls [38, 39] have also been reported more recently. The method works for proteins as large as the 723-residue malate synthase G (MSG) [40], which has a tumbling time of ~37 ns at 37 °C.

3.2 Mutagenesis

For larger proteins with molecular weights exceeding 100 kDa, the backbone assignment is not usually available; in fact it is often not easy to record a 2D amide TROSY [41] in these supralarge systems. For these proteins, alternative assignment strategies are required. One very useful method is to mutate those residues of interest and locate the changes in spectra: usually the disappeared peaks correspond to the residues mutated. Relatively conservative mutations are usually used to minimize the impact on local structure; for example, Ile can be mutated to Leu. This helped to assign two Ile (I149 and I151) δ1 methyl peaks of Clp protease, a 300 kDa tetradecameric complex, which was then used to delineate the dynamic nature of a key region for product release [41]. Mutation was also used in many other projects [3, 13, 33, 42, 43] to help establish the assignments. However, sometimes mutations lead to complex changes in the spectra, which make it very difficult to correlate the mutated residues with particular peaks.

3.3 The Art of Dissecting Large Proteins

Currently the most successful strategy relies on the assignments from smaller building blocks [3, 13]. In the case of the 20S proteasome core particle, which is a 670 kDa complex containing 28 subunits ($\alpha_7\beta_7\beta_7\alpha_7$), assignments were first made on a monomeric form of the α-protein. Many of these assignments were then transferred to the 360 kDa $\alpha_7\alpha_7$ complex (the correlation time is 120 ns at 50 °C), when the chemical shifts for methyl ^{1}H and ^{13}C as well as another two side chain ^{13}C (obtained using experiments similar to aforementioned out-and-back COSY) correlated well in the two forms. Additional assignments were made based on NOE correlations with these already assigned methyls, assuming proximity in the available crystal structure. Finally, methyl assignments from $\alpha_7\alpha_7$ were transferred to $\alpha_7\beta_7\beta_7\alpha_7$ in a similar way (Fig. 5).

Another successful example of this ‘dissection’ strategy is SecA, a 204 kDa ATPase from the Sec translocase. Unlike the proteasome, which is formed from relatively simple monomers, SecA is a dimer of very long chains (901 residues).

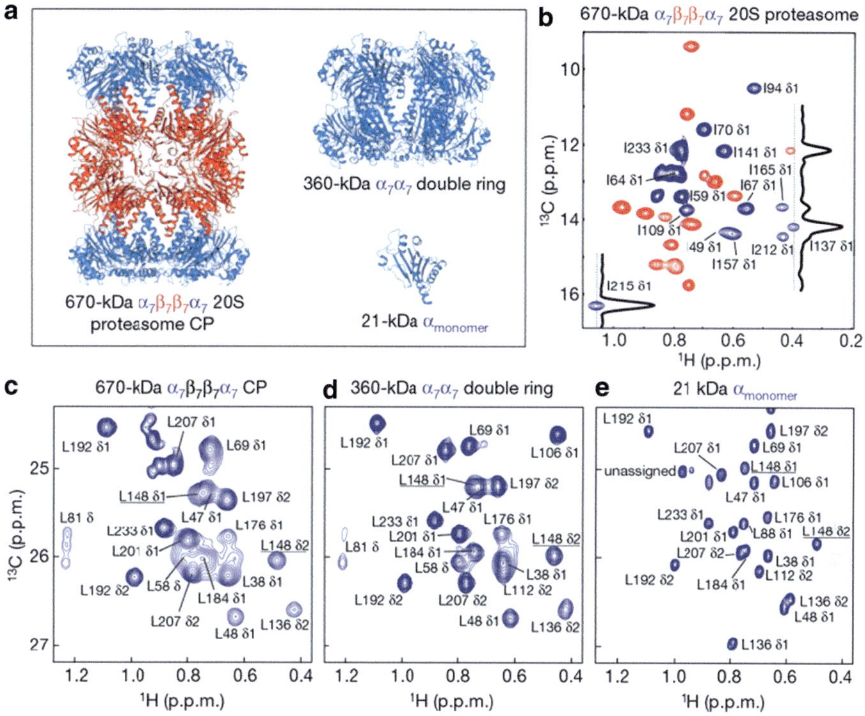

Fig. 5 (**a**) Ribbon diagrams of the 20S core particle (*CP*) ($\alpha_7\beta_7\beta_7\alpha_7$), the $\alpha_7\alpha_7$ double ring and the $\alpha_{monomer}$. (**b–e**), Methyl-TROSY spectra (800 MHz) of the CP (65 °C), $\alpha_7\alpha_7$ and $\alpha_{monomer}$ (50 °C). Correlations belonging to residues within α- or β-subunits are colour-coded *blue* or *red*, respectively. Apart from the spectrum in (**b**), only the α-subunit is NMR-active. The position of L148, one of the starting points in the assignment of the proteasome, is indicated by *underlining*. The correlations in (**c**), indicated by *asterisks*, are not present in all preparations and probably derive from impurities. Adapted from [3] with approval

Assignments were made step by step for constructs containing only one domain (residues 220–379), then two domains (residues 1–420), three domains (residues 1–610), and finally the full-length SecA (Fig. 6). Although this strategy represents a standard way to assign most of the observed methyl peaks confidently in a very large protein system, the process can be very laborious because it takes time to find suitable fragments (especially when mutagenesis is needed) and the assignment transfer has to be carried out cautiously [42].

3.4 Using Exchange Spectra and Pseudocontact Shifts

The above 'divide-and-conquer' approach was also applied to the 300 kDa aspartate transcarbamoylase (ATCase), containing six regulatory (r) chains and six catalytic (c) chains [42]. A 30 kDa r_2 dimer was used as the smaller block and

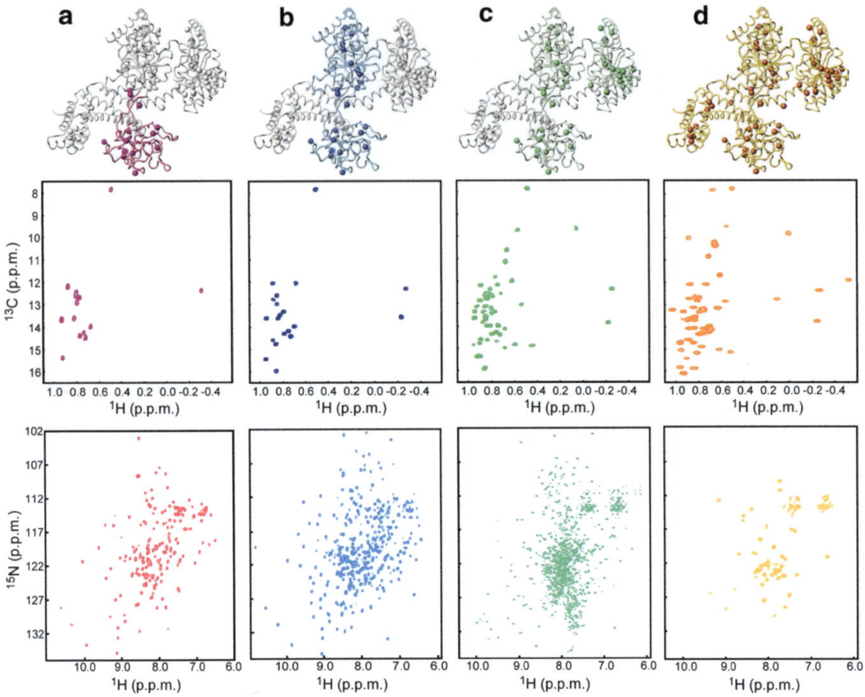

Fig. 6 Strategy for the assignment of methyl correlations of SecA. Each column displays a structural model of one of the protomers of SecA, with the domain or fragment studied in isolation being *highlighted*, along with the corresponding ^1H-^{13}C HMQC of Ile-δ1 methyls (displayed as *spheres* in the model) and the backbone ^1H-^{15}N HSQC. (**a**) PBD domain (residues 220–379), (**b**) SecAΔC/ΔIRA2 (residues 1–420, comprising NBD and PBD domains), (**c**) SecAΔC (residues 1–610, comprising NBD, PBD, and IRA2 domains), and (**d**) full-length SecA (residues 1–901). Only few resonances for the backbone of the full-length SecA are visible. Adapted from [13] with approval

assignments were obtained conventionally. Approximately 60% observed methyl peaks for the complex (only r-chains labelled) could be assigned with reference to the dimer. Two additional approaches were used to make more assignments. One involved mutagenesis on the interface of r_2 and c_3 to accelerate the exchange between r_2 in the free form and in the complex so that exchange cross-peaks could be observed, making the assignment transfer from free form to complex form unambiguous. However, exchange peaks cannot be observed unambiguously if there is degeneracy in one of the methyl chemical shifts or if the peaks are located in crowded regions. In addition, some peaks assigned in the mutated complex may not be easily transferred back to the wild complex due to shifted peak positions, which limits the usefulness of this approach. The other approach is to introduce an unpaired electron (such as using a Co^{2+} to replace Zn^{2+}) into the system, which causes pseudocontact shifts (PCSs) [44]. PCSs can be calculated according to the structure if the susceptibility tensor is known, or the susceptibility tensor can be established using a group of PCSs for assigned methyl groups. By comparison of measured PCS values to predicted values, ambiguity in the assignments using former methods could be solved, leading to more assignments, including some stereospecific ones. For ATCase, 86% peaks were assigned when all these methods were used.

3.5 Automatic Assignments

NOE spectroscopy (NOESY) spectra play a very important role in the assignment of many methyl resonances, with the help of available high resolution structures. Although NOESY is usually used to extend the initial assignments made through mutation or comparison with smaller fragments, it is possible to make assignments without resorting to these additional data [45], which reduced the time and cost. Both chemical shifts [46, 47] and NOE correlations can be predicted according to the structures, which are compared with the observed chemical shifts and NOE patterns of the methyls. An initial assignment can be made through this comparison, although only some very characteristic peaks could be confidently assigned at this stage. The consistency of the NOE network with the structure is then checked based on the assignments, which is subject to a systematic swap aiming at improving the consistency. In ideal cases such as the $\alpha_7\alpha_7$ complex, where overlap among methyl peaks is minimal and NOESY spectra are of very high quality and provide an extensive NOE network among the methyls, almost all the observed methyl peaks can be assigned correctly [45]. As the process is automated without any manual intervention, it takes only a few minutes. Even when the NOE network is relatively sparse, automatic optimization of the match between the methyl chemical shifts and NOE correlations with the structure can help to assign some of the peaks quickly. With more methyl groups other than the traditional ILV groups being labelled (see Sect. 2), it is anticipated that this strategy will be more useful as more NOE connections are detectable.

Using NOE or PCS for methyl assignments relies on high-resolution crystal structures, which might not be available or there might be discrepancies between the crystal structures and the structures in solution. Mutagenesis does not rely on available structures, but may lead to structural changes itself. A possible future development for methyl assignment in large proteins is site-specific labelling, which can simplify the spectra significantly without changing the protein structure and lead to confident assignments. In principle, the redundancy of genetic codons can be used so that the positions meant to be labelled use different codons from other positions of the same amino acid type in the template. One proposal has been published [48], although it may take years before we see it realized.

4 Applications of Methyl-TROSY in Large Proteins

With methyl-labelled samples, methyl-TROSY can provide detailed information regarding the structure, site-specific interaction, and dynamics of many important molecules that used to be too large for conventional NMR approaches. Spectra of good quality have been published for proteins with aggregate molecular weights of several hundred kiloDaltons, such as p97 [28] (Fig. 4), TET2 [49] and the CheA-CheW complex [50]. Here, we will focus on three systems that have been most extensively studied and represent the most successful cases to date.

4.1 Aspartate Transcarbamoylase

ATCase is known to adopt a pair of conformations, termed T and R. Using methyl-TROSY, Velyvis et al. showed that active site ligands (carbamoyl phosphate, CbmP) and analogues shifted the equilibrium from T to R (Fig. 7); shifts of the equilibrium between T and R states were also observed when nucleotides such as ATP and CTP were added. The NMR results supported the Monod–Wyman– Changeux (MWC) model to explain the cooperative binding and allosteric properties of ATCase [51, 52]. Interestingly, these were obtained even before the time-consuming methyl resonance assignment was complete. With assignments available, they were able to locate structural perturbations in the N-terminal tails of the regulatory chain by nucleotide binding, providing insights into the nature of the allosteric effect [42].

4.2 SecA

The potential of NMR in the structure determination of very large molecules is demonstrated by the study of a complex formed by SecA and a signal peptide [13]. Distance restraints were obtained using site-directed spin labelling (SDSL) on the peptide, causing distance-dependent broadening of the methyl resonances of SecA. Such paramagnetic relaxation enhancement (PRE) affected sites as far as ~30 Å away, therefore a large number of distance restraints were obtained. Spin labels at different positions in the peptide helped to place the peptide correctly onto SecA, while the hydrophobic region of the peptide itself was found to form an α-helix using transferred NOE spectroscopy (trNOESY). The structure helped to understand how SecA could recognize a diverse set of signal sequences. The binding groove on SecA was found to be partly occluded by the C-tail in the X-ray structure, and removal of the C-tail resulted in stronger binding. The authors also concluded that SecA interconverted between an open and closed conformation in solution according to NOE and PRE evidence (Fig. 8).

4.3 20S Proteasome

The 20S proteasome from *Thermoplasma acidophilum* was the first supramolecular system for which complete assignments of methyl peaks were obtained, and a series of studies have been carried out on it using NMR. Using $^{13}CHD_2$-labelled samples,

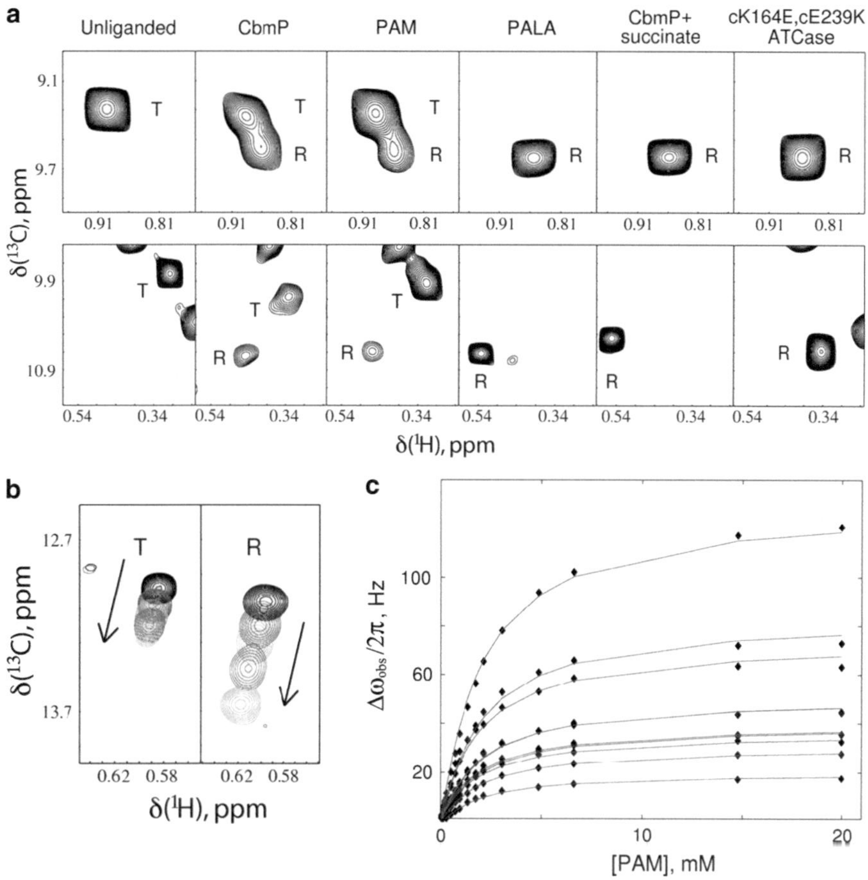

Fig. 7 Effects of binding of substrate and analogues on the spectrum of ATCase. (**a**) Observation of R conformation on ligand binding. Portions of methyl-TROSY spectra of U-[^{2}H] Ile-[$\delta 1^{13}C^1H_3$]-labelled ATCase (800 MHz, 37 °C) are shown that include a correlation derived from an Ile residue from the r-chain (*upper*) and from the c-chain (*lower*). *Column 1*, unliganded wild-type ATCase; *column 2*, wild-type ATCase saturated with 15 equiv. of carbamoyl phosphate (*CbmP*), i.e. ratio [CbmP]/[ATCase]$_{monomer}$ = 15; *column 3*, wild-type ATCase saturated with 58 equiv. of phosphonoacetamide (*PAM*); *column 4*, wild-type ATCase saturated with 1.5 equiv. of phosphonoacetyl-L-aspartate (*PALA*); *column 5*, wild-type ATCase saturated with 30 equiv. of CbmP and 75 equiv. of succinate; *column 6*, unliganded double mutant cK164E, cE239K–ATCase. *T* and *R* label peaks derived from ATCase in either the T or R conformation, respectively. (**b**) Titration of a correlation derived from the T state of wild-type ATCase (*left*) and from the R state (*right*) with PAM. Overlay of spectra with PAM concentrations ranging from 0 (*top*) to saturating (*bottom*). The direction of peak shift is indicated by an *arrow* in each case. Only three intermediate PAM concentrations are shown for clarity. (**c**) Global fitting of all shifting peaks from the titration of cK164E, cE239K–ATCase (R state) to a binding model with a single K_D. Reprinted from [51]

the side-chain dynamics at the pico- to nanosecond timescale was studied [3]. Values of order parameters squared (S^2) derived independently from ^{2}H and ^{13}C measurements were found to have excellent correlation, making the results highly

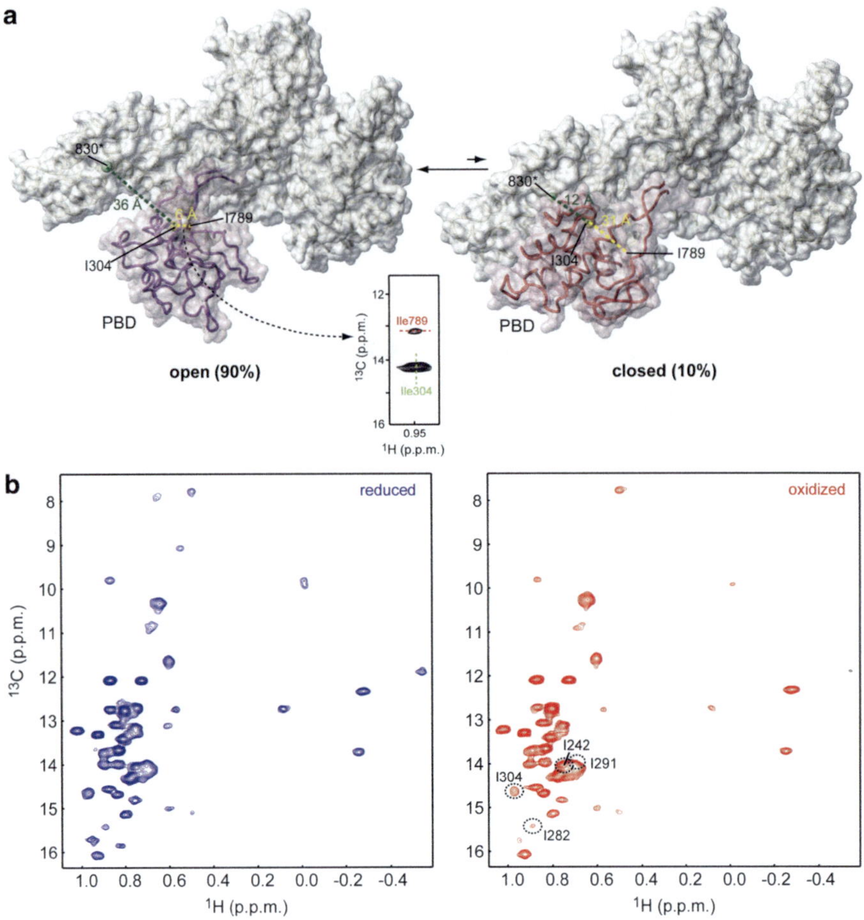

Fig. 8 SecA interconverts between an open and closed conformation in solution (**a**) SecA shown in the so-called open (*left*) and closed (*right*) conformations. Interconversion between the two conformations requires that the preprotein binding domain (*PBD*) undergoes a ~60° rigid-body rotation. PBD is displayed as *semitransparent* surface. The *spheres* indicates residue 830, in which a paramagnetic spin label was introduced, and residues Ile304 and Ile789. Characteristic distances in the two conformations are indicated. A strong NOE between Ile304 and Ile789 was observed, demonstrating that SecA adopts predominantly the open conformation in solution. (**b**) Overlaid ^{1}H-^{13}C HMQC spectra of SecA bearing a spin label in position 830 in the reduced (*left*) and oxidized (*right*) state. Residues that approach the spin label, even transiently, experience a broadening effect, which is suppressed in the reduced state. Adapted from [13] with permission

reliable despite the large size of the $\alpha_7\alpha_7$ complex (360 kDa). Most residues have similar S^2 values in $\alpha_7\beta_7\beta_7\alpha_7$, except for some residues at the α, β interface that show less flexibility. A cluster of flexible residues were identified on the outside of the barrel, which might carry a potential nuclear localization signal. Using relaxation dispersion experiments [53], the dynamics at the micro- to millisecond timescale was also studied. A few residues clustered on the inside of the antechamber were found to undergo structural exchanges, which might facilitate the transport of substrates to the sites of proteolysis. In addition, the N-terminal residues were also found to be mobile on the millisecond timescale and might fold into the antechamber and block the entrance. Titration with the 11S activator complex from *Trypanosoma brucei* to the core particle revealed the binding site, which was consistent with crystallography structure. Although the molecular weight of the complex 11S-$\alpha_7\beta_7\beta_7\alpha_7$-11S reached 1.1 MDa, the methyl-TROSY spectra were of good quality (see Fig. 9) and the dissociation constant could be readily obtained by fitting the peak intensities.

The proteasome is a target for anticancer drugs. Methyl-TROSY was used to study the binding of the antimalaria drug chloroquine with the 670 kDa 20S proteasome [54]. Unlike the previously reported inhibitors (such as MG132) of the proteasome, which bound to the catalytic region, chloroquine was shown to interact with regions between the α- and β-subunits, which is 20 Å away from the proteolytic active site. In fact, chloroquine could bind to proteasome simultaneously with MG132, indicating that the binding pockets were completely separated and that chloroquine represented a novel class of proteasome inhibitor.

The gating of archaeal 20S proteasome was further studied using $^{13}CH_3$-methionine-labelled samples [33]. There are two methionine residues (M1 and M6) in the N-terminal region, and another methionine (M-1) was introduced to augment them. Multiple peaks were observed in methyl-TROSY spectra, which were then assigned via mutagenesis. Both M1 and M-1 generated three peaks, originated from a major state (A) and two minor states (B and C) (see Fig. 10). PRE analysis suggested that the major state corresponded to a conformation with the N-terminal being outside of the barrel, while the minor states corresponded to conformations with the N-terminal extended inside the barrel. The in and out states of the N-termini were shown to interconvert on a timescale of seconds, based on magnetization exchange spectroscopy.

Recently, the substrate state inside the proteasome antechamber was studied [43]. ILV-methyl-labelled substrates were tethered to deuterated $\alpha_7\alpha_7$ to form stable substrate–proteasome complexes. Although all the substrates were well folded and generated well-resolved spectra at 50 °C in the free state, their HMQC spectra at the same temperature only contained clusters of broad peaks derived from each methyl type when they were tethered to positions inside the antechamber, suggesting that they were in unstructured states. The structures were not affected when they were tethered to the N-terminus or the outside of the antechamber. Some substrates became folded at lower temperatures, although they seemed to exist in several 'native-like' states because each methyl produced clustered peaks at the locations for native state proteins (Fig. 11). The interconversion between folded and unfolded states was slowed down when the proteins were encapsulated into the proteasome, and the stability was also drastically reduced. NMR relaxation experiments showed that the substrate proteins had similar overall tumbling times as the proteasome when encapsulated, suggesting that they interacted with the walls of the cavity. Such interactions should be of nonspecific nature because (1) the encapsulation only caused very small changes in the chemical shifts and relaxation

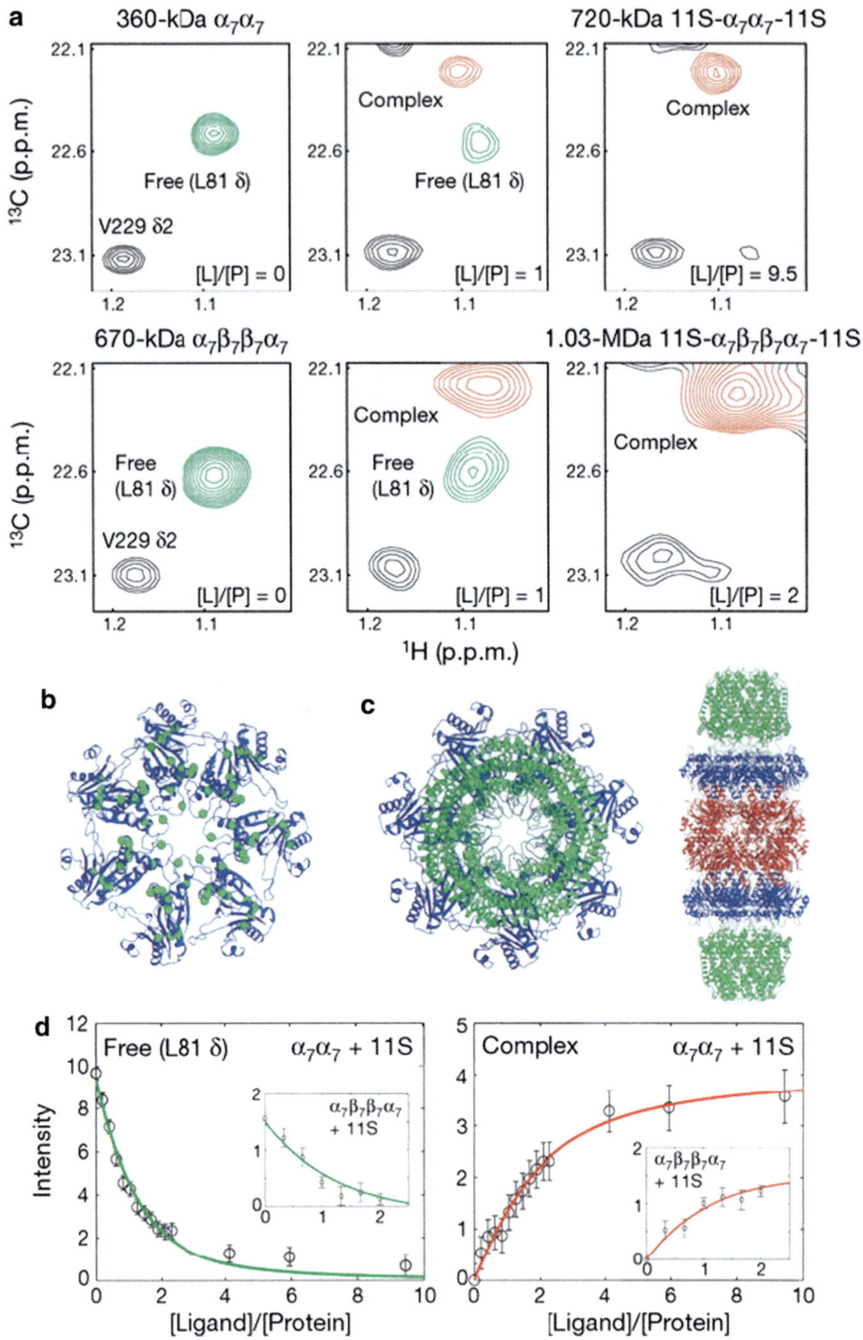

Fig. 9 (**a**) Methyl-TROSY spectra of $\alpha_7\alpha_7$ (*top*) and the core particle (*bottom*) during addition of the 11S activator complex. The complex is in slow exchange on the NMR chemical shift timescale;

properties of the α-subunit methyl groups; and (2) nitroxide spin labels on the substrate protein produced nearly identical decreases in peak intensities for those methyl groups located within 20 Å of the inside surface of the cavity. The ability to maintain the substrates in an unfolded state by the antechamber is important for the subsequent degradation.

5 Perspectives

Advances in methyl labelling techniques and the use of methyl-TROSY experiments have greatly expanded the range of molecular targets suitable for solution NMR studies [55]. Although it is not realistic to build high-resolution structures for very large proteins, NMR has been useful for providing new insights into important, unanswered biological questions. For example, the structure may have been solved by X-ray crystallography, but NMR can give complementary information on dynamics as well as structures not seen in the crystal.

Assignment of the observed signals is a major bottleneck for large proteins, although sometimes this can be bypassed. Mutagenesis can be used for rapid assignment of a few residues, while the complete assignment usually relies on pre-determined structures and spectra of smaller building blocks. Automation of the assignment is possible in favourable cases. We anticipate that the future development of site-specific labelling will provide a simple and fast way for achieving complete assignments without knowledge of structures.

Several methods have been developed to probe the structure and dynamics of large proteins. For example, RDC can be measured for methyl resonances [31], and relaxation dispersion protocols can be extracted in different ways. PRE measurements have been used to generate structural restraints for SecA complex and to decide the substrate location in the proteasome. Compared with the NOE, which relies on recording 3D experiments, PRE is easier to detect and can explore much longer distances. This proves very useful for large proteins because the spectral sensitivity is often limited by low concentrations or conformational exchange problems. Although some proteins generate exceptional spectra, more often one has to work with lower quality spectra and, in these cases, new tools such as PRE are extremely useful.

Fig. 9 (continued) assignments of correlations from the complex are not available. (**b**) Residues whose resonances are affected by the 11S interaction are mapped on the proteasome structure. (**c**) Crystal structure of the 11S-proteasome complex. (**d**) The intensities of resonances during the titration are used to obtain an approximate dissociation constant ($K_D = 12 \pm 10\ \mu M$) for the $\alpha_7\alpha_7$/11S interaction that is consistent with the core particle titration data (see *insets*). The decrease in intensity of one of the correlations from L81 is shown on the *left* and the concomitant increase in a 'bound' peak on the *right*. Errors are quantified from signal-to-noise in spectra. [*Ligand*] and [*Protein*] refer to total ligand and protein concentrations. Adapted from [3] with permission

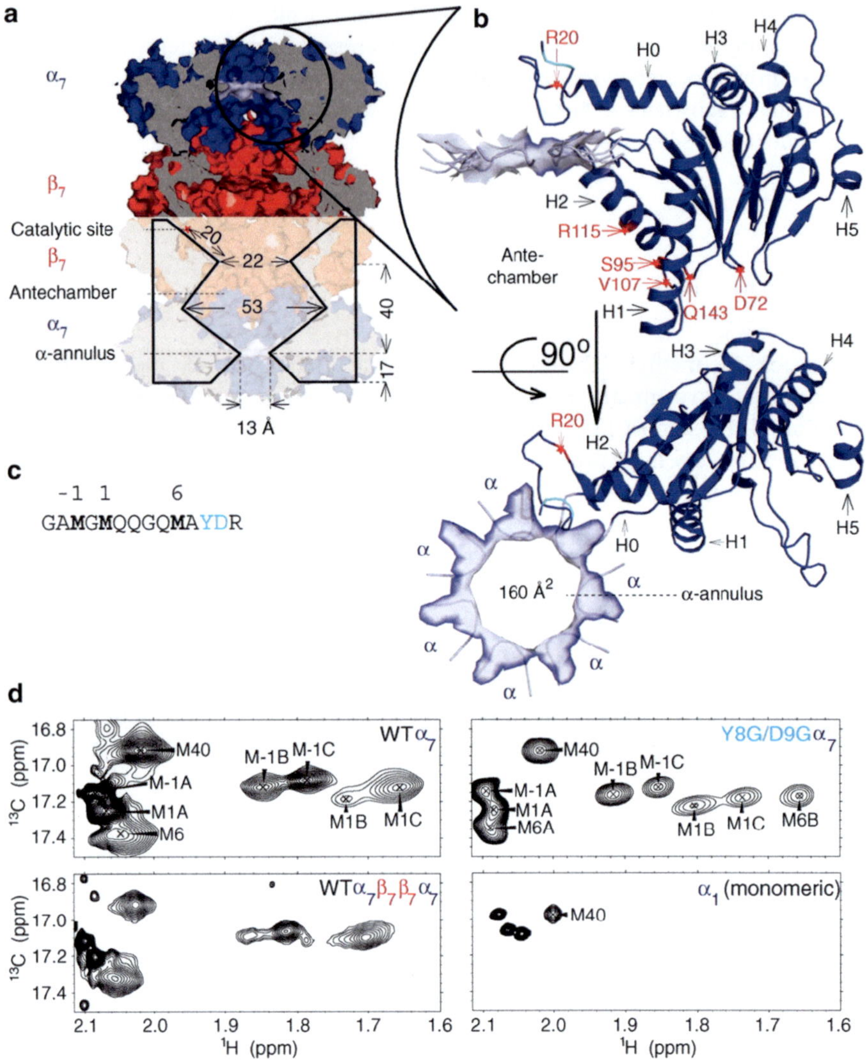

Fig. 10 Multiple conformations for the proteasome gating residues. (**a**) Space-filling, cross-sectional side-view representation of the 20*S* core-particle proteasome ($\alpha_7\beta_7\beta_7\alpha_7$) from *Thermoplasma acidophilum*, showing the relative positions of the α-annulus, the antechambers, and the catalytic chamber, as well as the dimensions of the molecule [PDB accession code 1YA7]. (**b**) Each α-ring is composed of seven identical α-subunits. A single α-subunit is shown (two orientations), along with the α-annulus that is composed of residues from all seven subunits. The locations of added spin labels for PRE measurements are indicated. (**c**) Primary sequence of the N-terminal α-subunit residues, including GAMG introduced by cloning. The Met residues used as probes are shown in *bold*. When YD residues (*grey*) are mutated to G residues, the out conformation of the gating residues is destabilized. (**d**) ^{1}H-^{13}C Met methyl HMQC spectra of the wild-type (*WT*) α_7 ring, WT $\alpha_7\beta_7\beta_7\alpha_7$, and the Y8G/D9G α_7 ring showing multiple correlations for M1, M-1 (denoted by *A*, *B*, and *C*), and M6 (Y8G/D9G α_7 ring, denoted by *A* and *B*). Multiple correlations are not observed for the isolated α-subunit (*bottom right*). In all spectra, the M120 correlation is outside the spectral window shown here. Adapted from [33] with permission

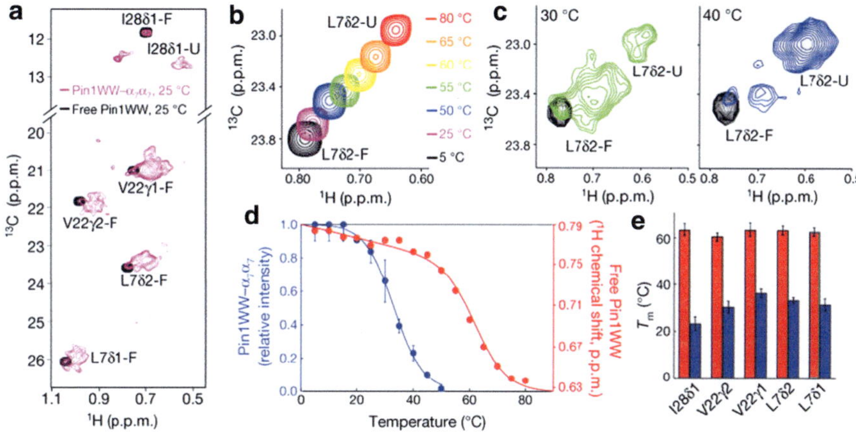

Fig. 11 (**a**) Spectrum of ILV-methyl-labelled Pin1WW in $\alpha_7\alpha_7$ at 25 °C (*magenta*). Free Pin1WW is shown in *black* as a reference. (**b**, **c**) Spectral regions showing the temperature dependence of the L7δ2 methyl peak for Pin1WW free in solution (**b**) and tethered to $\alpha_7\alpha_7$ (**c**). (**d**) Representative melting curves for L7δ2 of free (*red*) and $\alpha_7\alpha_7$-encapsulated (*blue*) Pin1WW, derived using either chemical shifts (free Pin1WW) or the relative peak intensities of unfolded and folded states (encapsulated Pin1WW). Fits to standard equations for two-state melting are shown (*solid lines*). (**e**) Melting temperature (T_m) values derived from fits for each of the methyl groups for free (*red*) and encapsulated (*blue*) Pin1WW. For Pin1WW–$\alpha_7\alpha_7$, *error bars* are based on measurements of three separate samples. For free Pin1WW, *error bars* in chemical shifts, based on triplicate measurements (<0.3 p.p.b. for ^{13}C and <0.1 p.p.b. for ^{1}H), are too small to be visible relative to the size of the marker. Adapted from [43] with permission from Macmillan Publishers Ltd., copyright 2010

In addition to the examples mentioned in Sect. 4, methyl-TROSY has also been used to explore co-translational folding of a ribosome-bound nascent chain [56] as well as the structural basis for ion channel gating [57]. It is worth mentioning that most membrane proteins fall into the size range of TROSY experiments, and the methods reviewed here are relevant to their study.

References

1. Kay LE, Gardner KH (1997) Curr Opin Struct Biol 7:722
2. Xu YQ, Zheng Y, Fan JS, Yang DW (2006) Nat Methods 3:931
3. Sprangers R, Kay LE (2007) Nature 445:618
4. Lemaster DM, Richards FM (1988) Biochemistry 27:142
5. Pervushin K, Riek R, Wider G, Wuthrich K (1997) Proc Natl Acad Sci USA 94:12366
6. Keeler J (2005) Understanding NMR spectroscopy. Wiley, New York
7. Salzmann M, Pervushin K, Wider G, Senn H, Wuthrich K (1998) Proc Natl Acad Sci USA 95:13585
8. Fiaux J, Bertelsen EB, Horwich AL, Wuthrich K (2002) Nature 418:207
9. Pervushin K (2000) Q Rev Biophys 33:161

10. Salzmann M, Pervushin K, Wider G, Senn H, Wuthrich K (2000) J Am Chem Soc 122:7543
11. Pervushin K, Riek R, Wider G, Wuthrich K (1998) J Am Chem Soc 120:6394
12. Tugarinov V, Hwang PM, Ollerenshaw JE, Kay LE (2003) J Am Chem Soc 125:10420
13. Gelis I, Bonvin A, Keramisanou D, Koukaki M, Gouridis G, Karamanou S, Economou A, Kalodimos CG (2007) Cell 131:756
14. Tugarinov V, Hwang PM, Kay LE (2004) Annu Rev Biochem 73:107
15. Sprangers R, Velyvis A, Kay LE (2007) Nat Methods 4:697
16. Tugarinov V, Kanelis V, Kay LE (2006) Nat Protoc 1:749
17. Sivashanmugam A, Murray V, Cui CX, Zhang YH, Wang JJ, Li QQ (2009) Protein Sci 18: 936
18. Rosen MK, Gardner KH, Willis RC, Parris WE, Pawson T, Kay LE (1996) J Mol Biol 263:627
19. Otten R, Chu B, Krewulak KD, Vogel HJ, Mulder FAA (2010) J Am Chem Soc 132:2952
20. Gardner KH, Kay LE (1998) Annu Rev Biophys Biomol Struct 27:357
21. Gardner KH, Kay LE (1997) J Am Chem Soc 119:7599
22. Goto NK, Gardner KH, Mueller GA, Willis RC, Kay LE (1999) J Biomol NMR 13:369
23. Tugarinov V, Kay LE (2004) J Biomol NMR 28:165
24. Gans P, Hamelin O, Sounier R, Ayala I, Dura MA, Amero CD, Noirclerc-Savoye M, Franzetti B, Plevin MJ, Boisbouvier J (2010) Angew Chem Int Ed 49:1958
25. Ruschak AM, Velyvis A, Kay LE (2010) J Biomol NMR 48:129
26. Baldwin AJ, Religa TL, Hansen DF, Bouvignies G, Kay LE (2010) J Am Chem Soc 132:10992
27. Religa TL, Kay LE (2010) J Biomol NMR 47:163
28. Isaacson RL, Simpson PJ, Liu M, Cota E, Zhang X, Freemont P, Matthews S (2007) J Am Chem Soc 129:15428
29. Ayala I, Sounier R, Use N, Gans P, Boisbouvier J (2009) J Biomol NMR 43:111
30. Godoy-Ruiz R, Guo C, Tugarinov V (2010) J Am Chem Soc 132:18340
31. Sprangers R, Kay LE (2007) J Am Chem Soc 129:12668
32. Fischer M, Kloiber K, Hausler J, Ledolter K, Konrat R, Schmid W (2007) ChemBioChem 8:610
33. Religa TL, Sprangers R, Kay LE (2010) Science 328:98
34. Kigawa T, Yabuki T, Yoshida Y, Tsutsui M, Ito Y, Shibata T, Yokoyama S (1999) FEBS Lett 442:15
35. Kainosho M, Torizawa T, Iwashita Y, Terauchi T, Mei Ono A, Guntert P (2006) Nature 440:52
36. Tugarinov V, Kay LE (2003) J Am Chem Soc 125:13868
37. Tugarinov V, Kay LE (2003) J Am Chem Soc 125:5701
38. Sheppard D, Guo CY, Tugarinov V (2009) J Am Chem Soc 131:1364
39. Sheppard D, Guo CY, Tugarinov V (2009) J Biomol NMR 43:229
40. Tugarinov V, Choy WY, Orekhov VY, Kay LE (2005) Proc Natl Acad Sci USA 102:622
41. Sprangers R, Gribun A, Hwang PM, Houry WA, Kay LE (2005) Proc Natl Acad Sci USA 102:16678
42. Velyvis A, Schachman HK, Kay LE (2009) J Am Chem Soc 131:16534
43. Ruschak AM, Religa TL, Breuer S, Witt S, Kay LE (2010) Nature 467:868
44. John M, Schmitz C, Park AY, Dixon NE, Huber T, Otting G (2007) J Am Chem Soc 129:13749
45. Xu YQ, Liu MH, Simpson PJ, Isaacson R, Cota E, Marchant J, Yang DW, Zhang XD, Freemont P, Matthews S (2009) J Am Chem Soc 131:9480
46. Neal S, Nip AM, Zhang HY, Wishart DS (2003) J Biomol NMR 26:215
47. Osapay K, Case DA (1991) J Am Chem Soc 113:9436
48. Gaspari Z, Pal G, Perczel A (2008) Bioessays 30:772
49. Amero C, Schanda P, Dura MA, Ayala I, Marion D, Franzetti B, Brutscher B, Boisbouvier J (2009) J Am Chem Soc 131:3448
50. Hamel DJ, Dahlquist FW (2005) J Am Chem Soc 127:9676
51. Velyvis A, Yang YR, Schachman HK, Kay LE (2007) Proc Natl Acad Sci USA 104:8815

52. Velyvis A, Schachman HK, Kay LE (2009) J Mol Biol 387:540
53. Korzhnev DM, Kloiber K, Kanelis V, Tugarinov V, Kay LE (2004) J Am Chem Soc 126: 3964
54. Sprangers R, Li XM, Mao XL, Rubinstein JL, Schimmer AD, Kay LE (2008) Biochemistry 47:6727
55. Ruschak AM, Kay LE (2009) J Biomol NMR 46:75
56. Hsu ST, Cabrita LD, Fucini P, Christodoulou J, Dobson CM (2009) J Am Chem Soc 131:8366
57. Imai S, Osawa M, Takeuchi K, Shimada I (2010) Proc Natl Acad Sci USA 107:6216

Top Curr Chem (2013) 335: 121–156
DOI: 10.1007/128_2012_417

Published online: 16 March 2013

Solid-State NMR Spectroscopy of Proteins

Henrik Müller, Manuel Etzkorn, and Henrike Heise

Abstract Solid-state NMR spectroscopy proved to be a versatile tool for characterization of structure and dynamics of complex biochemical systems. In particular, magic angle spinning (MAS) solid-state NMR came to maturity for application towards structural elucidation of biological macromolecules. Current challenges in applying solid-state NMR as well as progress achieved recently will be discussed in the following chapter focusing on conceptual aspects important for structural elucidation of proteins.

Keywords Amyloid fibrils · Biomolecular solid-state NMR spectroscopy · Dynamic nuclear polarization · Membrane proteins

Contents

To my father Werner Heise (1944–2013)

H. Müller, M. Etzkorn, and H. Heise (✉)
Institute of Physical Biology Heinrich-Heine-University of Düsseldorf, 40225 Düsseldorf, Germany

Institute of Complex Systems, Structural Biochemistry (ICS-6), Research Center Jülich, 52425 Jülich, Germany
e-mail: h.heise@fz-juelich.de

Abbreviations

APP	Amyloid precursor protein
ASR	Anabaena sensory rhodopsin
CE	Cross-effect
CRAMPS	Combined rotation and multiple pulse spectroscopy
CW	Continuous wave
DARR	Dipolar assisted rotational resonance
DNP	Dynamic nuclear polarization
EPR	Electron paramagnetic resonance
FSLG	Frequency-switched Lee-Goldburg
GPCR	G protein-coupled receptor
HORROR	Homonuclear rotary resonance
INEPT	Insensitive nuclei enhanced by polarization transfer
MAS	Magic angle spinning
PDSD	Proton-driven spin diffusion
PMLG	Phase-modulated Lee-Goldburg
PRE	Paramagnetic relaxation enhancement
REDOR	Rotational echo double resonance
SE	Solid effect
TEDOR	Transferred echo double resonance
TM	Transmembrane helix

1 Introduction

In the last decade, solid-state NMR spectroscopy has proven its enormous potential for structural investigations of biological macromolecules or macromolecular complexes. In contrast to X-ray crystallography, which requires crystalline samples, and solution-state NMR spectroscopy, which requires fast tumbling molecules and thus molecular weights of less than a few hundred kDa, solid-state NMR spectroscopy imposes no restrictions upon solubility, crystallizability, molecular size, or even purity of the sample. Solid-state NMR spectroscopy is a particularly powerful method for investigating large protein assemblies or aggregates, integral membrane proteins reconstituted into lipid bilayers, biominerals, protein–DNA complexes, or even full virus particles.

In solution-state NMR spectroscopy, rapid isotropic tumbling leads to a complete averaging of chemical shift anisotropy, homonuclear and heteronuclear dipolar couplings, and also, in the case of nuclei with spin $>1/2$, of quadrupolar interactions. In contrast, in solids the resonance frequency and the magnitude of internuclear couplings depend on the orientation of the molecule within the magnetic field. These anisotropic interactions can either be exploited for information on dynamics and orientation, or overcome. This can be accomplished in different ways:

- If the molecules have no preferred orientation in the sample, signals are dominated by their chemical shift anisotropy or the quadrupolar interaction, which leads to a distinct powder pattern. Even in the absence of isotropic tumbling, however, molecules can rotate in a preferred direction resulting in a partial averaging of the powder pattern. Especially in lipid bilayers or liposomes, dynamics of membranes can be probed by the line shapes either in ^{31}P spectra of the lipid head groups or of ^{2}H spectra in selectively deuterated positions in lipid side chains.
- Membrane proteins can be oriented macroscopically, e.g. by reconstitution into lipid bilayers on glass plates which are aligned perpendicular to the magnetic field. Since all membrane-reconstituted proteins adopt the same orientation in lipid bilayers, narrow lines are observed, and their chemical shifts as well as their dipolar couplings are a signature of the orientation [1].
- Anisotropic interactions can be averaged out by magic angle spinning (MAS) NMR: [2] by rapidly spinning the sample around an axis inclined at an angle of 54.7° whereby dipolar couplings, chemical shift anisotropy, and first-order quadrupolar interactions are averaged to zero. In principle, line widths as small as in liquid-state NMR spectroscopy can be obtained (Fig. 1).

The rapid methodological development of solid-state NMR spectroscopy, in particular MAS NMR spectroscopy, during the last decade as well as more sophisticated labeling techniques for proteins have established the possibility of also applying multi-dimensional NMR spectroscopy to biological macromolecules and complex multi-component systems [3–5].

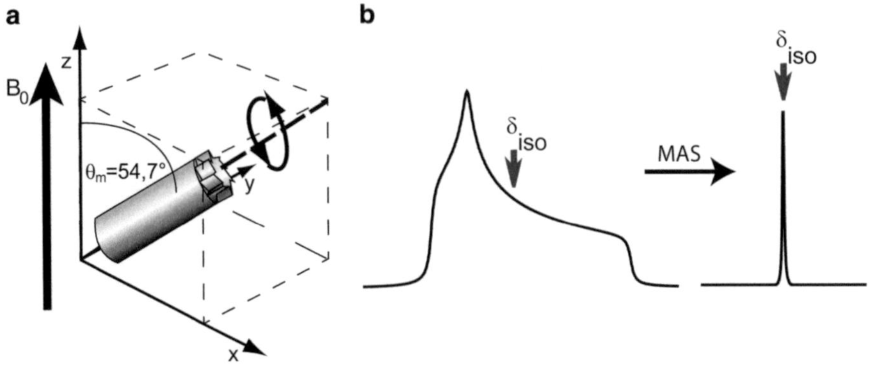

Fig. 1 Principle of magic angle spinning (MAS) solid state NMR. (**a**) Rotation of the sample rotor about the "magic angle" $\theta_m = 54.7°$ with respect to the magnetic field B_0, i.e., the space diagonal. (**b**) Typical solid state NMR spectrum without MAS (*left hand side*, simulated data) and with MAS (*right hand side*). The *arrow* indicates the isotropic chemical shift δ_{iso}

In the current chapter we focus on recent techniques and applications in MAS-NMR spectroscopy of proteins and biomolecules.

2 Sample Preparation

2.1 Isotope Labeling

Due to high sensitivity and resolution, ^{1}H NMR detection is the basis of most solution-state NMR spectra of proteins. Solid-state ^{1}H NMR spectra of proteins, however, are generally broadened beyond resolution due to spin relaxation in a network of dipolar coupled protons. Therefore, solid-state NMR spectroscopy of larger molecules like proteins critically depends on labeling with the stable isotopes ^{13}C, ^{15}N, and/or ^{2}H. Depending on the method chosen for protein production, a variety of isotope labeling strategies has been developed [6–9].

For peptides below a length of about 40 amino acids, solid phase synthesis can be applied in which nascent peptides are bound to a bead and synthesized gradually in iterated cycles of deviating chemical environments. Thus, site-specific introduction of isotope labels at any desired position is possible.

Larger proteins are usually obtained by recombinant expression in *E. coli*, yeast [10], insect cells [11], or mammalian cells [12]. In these cases, the isotope labeling pattern depends critically on the carbon- and nitrogen-containing precursors in the growth medium. For ^{13}C labeling, all intermediates of the glycolysis pathway and the citric acid cycle can serve as a carbon source. If the bacterial minimal growth medium contains glucose or its derivatives such as glycerol, pyruvate, or acetate as the sole carbon source and ammonium salts as the sole nitrogen source, uniformly ^{13}C and

^{15}N-labeled proteins can be obtained. Although such a uniform labeling pattern allows for collecting the maximum amount of spectroscopic information, the size of the protein as well as an unfortunate amino acid distribution might result in spectral overlap. In such cases, a reduction in the number of labeled amino acids might become crucial which can be achieved, e.g. by adding unlabeled amino acids to the growth medium. The success of this so-called reverse labeling depends on the metabolic pathway for the biosynthesis of the chosen amino acid [13, 14] which can be improved by the use of auxotrophic strains unable to synthesize the respective amino acids [15]. Complementary, type-selective amino acid labeling can be obtained by adding a labeled amino acid type to an otherwise unlabeled growth medium [16]. Atom position-specific labeling can be achieved by using reduced labeled precursors such as 1,3- or 2-^{13}C-glycerol or 1- or 2-^{13}C glucose resulting in characteristic ^{13}C isotope distributions within each amino acid type [6, 9, 14, 17]. Incorporation of isolated ^{13}C isotopes solely at methyl sites is obtained by using 3-^{13}C-pyruvate as sole carbon source [18].

An even more sophisticated approach comprises protein synthesis by cell-free expression: as the amino acids added in a cell-free system are less exposed to bacterial metabolism, efficiency and selectivity of incorporation of specific isotope-labeled amino acids are only marginally hampered by isotope scrambling often observed in *in vivo* expression systems [19]. Even non-canonical amino acids can be incorporated allowing for site-specific labeling of proteins [20, 21].

At least for non-crystalline samples, ^{1}H detection is still a major challenge. In solution NMR, the development of heteronuclear multidimensional NMR spectroscopy has paved the way for the assignment of backbone and side-chain resonances of proteins with molecular masses of over 30 kDa. This success has mainly been achieved through the production of highly deuterated samples because the replacement of protons with deuterons significantly improves the ^{1}H resolution [22–25]. Innovative labeling schemes, e.g. with 5 % protonation at non-exchangeable sites, enable ^{1}H-detection with high resolution resulting in determination of long range proton proton distances in solid-state MAS NMR spectroscopy [26]. However, protein deuteration is accompanied by several disadvantages. It not only depletes the number of ^{1}H–^{1}H distance restraints but also influences NMR resonance frequencies and cross polarization transfer efficiencies. In contrast, full protonation simplifies sample preparation and permits a more complete chemical shift assignment to be obtained from only one sample. Using the fully protonated 56-residue β1 immunoglobulin-binding domain of protein G (GB1) in a 1.6-mm rotor at 40 kHz MAS and 500 MHz proton frequency, proton line widths of 1 ppm and a sensitivity enhancement of up to four times compared to direct ^{13}C and ^{15}N detection could be observed. Apart from this, 3D pulse sequences transferring magnetization between heteronuclei such us CαNH, CONH, and NCαH even allowed full backbone and partial side-chain proton assignments [27].

A mixing of molecular species with different labeling patterns offers the possibility of obtaining intermolecular distance restraints [28].

Aiming at investigating a complex macromolecular surrounding such as whole cell preparations it has to be guaranteed that predominantly the protein of interest is isotope-labeled. In order to reduce the overall host protein content, single protein production, taking advantage of an mRNA interferase which cleaves all not genetically engineered mRNAs, constitutes a major milestone to a further improvement of sample quality [29].

2.2 *Microcrystals and Macromolecular Complexes*

The quality of solid-state NMR spectra critically depends on the properties of the sample. Depending on size, nature, and state of the protein, different ways of sample preparations can be used for solid-state NMR spectroscopy.

Pioneering solid-state NMR investigations on extensively isotope labeled proteins were performed on immobilized and often microcrystalline proteins. Especially for the later, well-resolved spectra could be obtained due to the absence of conformational disorder. Conformational ensembles of completely or partially unfolded proteins or multimers can be studied in frozen solution [30–32], where a lack of secondary structure does not result in an average random coil chemical shift but gives rise to a continuous distribution of chemical shifts [33]. Recently it was found that solid-state NMR spectroscopy can be performed on proteins >100 kDa even in solution due to transient sedimentation of large proteins at the rotor walls under fast MAS. Although these sediments are non-crystalline, the proteins are sufficiently immobilized suppressing motional averaging of dipolar interactions and chemical shifts [34, 35].

In contrast to solution-state NMR spectroscopy, the size of protein assemblies does not affect the line width in solid-state NMR spectroscopy. As only the number of inequivalent amino acids in a sample leads to spectral crowding and resonance overlap, large protein assemblies formed from one or only a few types of monomers are extremely suitable for solid-state NMR spectroscopy. Recent examples include a variety of amyloid fibrils [3, 36, 37], full virus particles [38, 39], and a secretion needle [40].

2.3 *Amyloid Fibrils: Seeded Versus Spontaneous Fibrillation*

A particular challenge associated with the study of assemblies of proteins in their non-native conformation is structural heterogeneity or polymorphism. As proteins may adopt different conformations upon aggregation, the structure of such fibrils can critically depend on the exact conditions of fibrillation. Minor variations of the pH, salt concentration [41], and even stirring of the solution [42] have been shown to have a tremendous influence on the morphology of the fibrils as well as the detailed molecular structure.

The fibrillation kinetics is characterized by a lag phase followed by a sigmoidal increase which is indicative of a nucleation-dependent aggregation, in which the formation of a nucleus is the rate-determining step. A lag phase can be circumvented by seeding with fibrils which have either been generated *in vitro*, or were isolated from *in vivo* material.

For amyloid fibrils of wild-type Aβ as well as a disease-related mutation thereof, sample homogeneity could be improved using a repeated seeding protocol [43, 44]. Increased homogeneity was ascribed to the selection of one fibril structure, which is kinetically as well as thermodynamically favored, out of a polymorphic mixture of different fibril types in the initial sample. However, the ultimate goal of this sort of study is the elucidation of disease-relevant conformations of the misfolded protein. As solid-state NMR studies rely on specific or uniform labeling with the NMR active isotopes ^{13}C and/or ^{15}N, structural investigations of amyloid fibrils purified from *in vivo* brain material – such as prion rods or fibrils generated from Alzheimer's plaques or Lewy bodies – are still impossible. In such cases, templating of isotope-labeled recombinantly expressed proteins with brain-derived seeds becomes the method of choice for investigating fibrils in a conformation as close to the *in vivo* state as possible. First promising attempts include Aβ [45] or the yeast prion Ure2p [46].

2.4 Membrane Proteins in a Native-Like Environment

In terms of sample complexity, the preparation and investigation of membrane proteins is particularly challenging because of their intrinsic hydrophobic nature and the necessity of using a suitable membrane-mimicking environment (Fig. 2) [47]. For solid-state NMR the reconstitution into lipid bilayers is often the method of choice. Although incorporation of detergent-solubilized membrane proteins into liposomes occurs spontaneously by decreasing the detergent concentration, it has to be demonstrated that reconstituted membrane proteins are fully functional under physiological conditions of pH, salinity, and the presence of endogenous ligands. As an advantageous model of biological membranes, so-called bicelles became popular during the last decade due to their capability to align in a magnetic field. They are composed of a mixture of long chain (14–18 carbon atoms) and short chain (6–8 carbon atoms) surfactants and hence represent an intermediate between lipid vesicles and micelles. Since bicelles are not disrupted by MAS, isotropic solid-state ^{13}C-, ^{15}N-, and ^{31}P-NMR spectroscopy can be applied for structural determination of molecules in membranes [48–51]. Finally, cellular membranes can be mimicked by discoidal nanoscale lipid bilayers, so-called nanodiscs, which are confined and stabilized by amphiphatic helical scaffold proteins [52]. They can be produced from a variety of lipids, are stable over a broad range of temperatures, and are accessible from both sides of the lipid bilayer. Since nanodiscs represent a more native environment than micelles or bicelles, they are an outstanding model system for understanding membrane protein function. Heteronuclear solid-state NMR

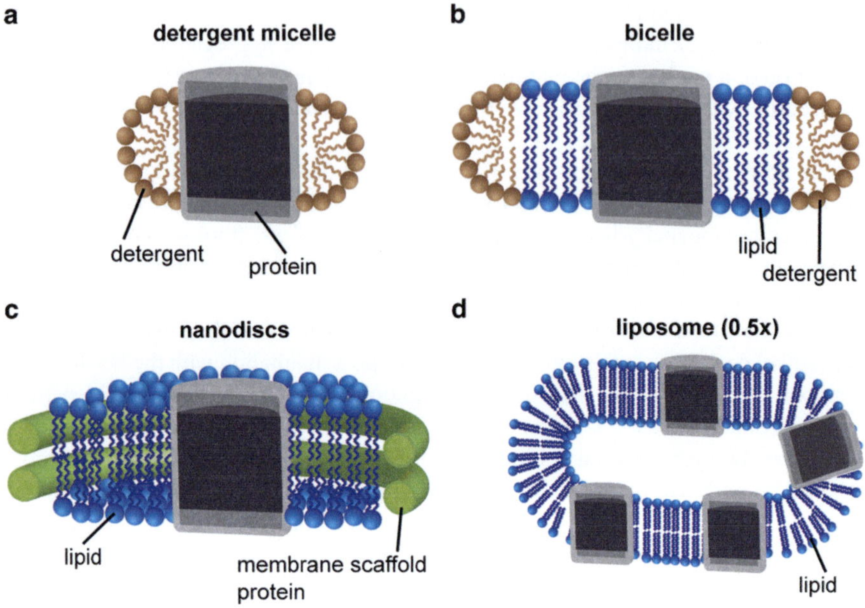

Fig. 2 Schematic representation of most common membrane-mimicking environments compatible with solution and/or solid-state NMR studies of integral membrane proteins. Micelle (**a**), bicelle (**b**), nanolipoprotein particles also known as nanodiscs (**c**), and liposomes (**d**). The membrane protein is depicted in *gray*, detergents are colored in *brown* and lipids in *blue*. The apolipoprotein also known as membrane scaffold protein is represented as a *green ring*. Liposomes are scaled down by a factor of 2

studies of nanodisc-incorporated proteins demonstrated the utility of MAS NMR to study the structure of high molecular weight lipid protein complexes [53–55].

3 Hardware Improvements: High Fields, Cryo-NMR, Fast MAS

One of the most obvious improvements during the last two decades of biomolecular NMR spectroscopy has been the development of high magnetic field strengths. As the chemical shift scales linearly with the external magnetic field, the spectral resolution increases linearly if the line widths are determined predominantly by the life time of the spin states (i.e. if the lines are homogeneously broadened as defined by Maricq and Waugh [56]). Furthermore, the signal intensity is proportional to the population difference between the two spin states of spin ½ nuclei given by the Boltzmann distribution. The higher the magnetic field, the higher the number of excitable spins in the low energy state resulting in an at least linear improvement of NMR-sensitivity as well. The strongest NMR instruments currently commercially available operate at a magnetic field strength of 23.5 T, corresponding to a proton Larmor frequency of 1 GHz.

The population difference given by the Boltzmann distribution can further be enhanced by lowering the sample temperature. Cryo-solid-state NMR at temperatures down to 90 K demonstrates that site-specific insights into molecular nature and dynamics of peptides and proteins can also be achieved in this temperature regime [57]. The increase in sensitivity, however, may be gained at the expense of spectral resolution which degrades substantially at temperatures below 210 K [58]. The observed line broadening reflects the freezing of molecular motions and side chain rotations into discrete populations. The dynamic transitions are reversible, at least in microcrystalline systems, excluding cold-denaturation or damage to the crystal lattice by freezing [57, 59, 60].

Recent developments in probe head technology have enabled MAS spinning rates of 70 kHz and more, which are particularly attractive in combination with direct proton detection [61, 62].

Finally, spin hyperpolarization techniques for obtaining spin polarizations exceeding the Boltzman limit by several orders of magnitude like optical polarization, para hydrogen induced hyperpolarization, or dynamic nuclear polarization (DNP) are subject of intense research. In the last section we will highlight recent break-throughs and latest contributions of DNP in the field of biomolecular solid-state NMR spectroscopy.

4 Basic Principles and Recent Pulse Sequence Improvements

4.1 Homogeneous Line Broadening and Proton Decoupling

Rapid exchange of energy quanta between nuclear spins in a 3D network by flip-flop transitions interferes with averaging by magic angle spinning, as the neighboring spins can change their spin state spontaneously and unpredictably during the rotor period. As a consequence rotational echoes decay fast and proton spectra remain homogeneously broadened even at ultra-high spinning speeds up to 70 kHz [56]. For high resolution, solid-state NMR spectroscopy of biological macromolecules depends on isotope labeling with ^{13}C and ^{15}N or on dilution of proton spins by extensive sample deuteration. Likewise, heteronuclear dipolar coupling of ^{13}C or ^{15}N spins to a network of strongly coupled proton spins prevents the refocusing of rare spins after each rotor period into a rotational echo. For high resolution in solid-state NMR spectroscopy, high power proton decoupling during evolution and detection periods is mandatory.

Homonuclear proton decoupling in the indirect evolution or in the direct detection period can be achieved by a net evolution of the protons around an effective field axis which is tilted by the magic angle of 54.7° from the magnetic field axis: in this interaction frame, the zero order term of the homonuclear dipolar Hamiltonian vanishes to zero. For indirect evolution periods this is achieved by irradiating with a strong rf field characterized by a resonance offset Ω that is matched to the rf field strength ω_{rf} according to $\omega_{rf}/\Omega = \sqrt{2}$ [63]. Since the magnetization does not precess freely but nutates around a tilted axis, the chemical shift as well as heteronuclear dipolar couplings are reduced by a factor of $1/\sqrt{3}$. First-order correction terms of the average homonuclear dipolar Hamiltonian are averaged out by either switching the resonance offset between $+\Omega$ and $-\Omega$ (frequency switched Lee-Goldburg, FSLG) [64]. Alternating frequency offsets can also be mimicked by continuous phase modulation of the rf pulse (phase-modulated Lee-Goldburg, PMLG [65] or DUMBO [66]).

For homonuclear decoupling during acquisition, a net evolution of the magnetization around an axis tilted by the magic angle from the B_0 field can be achieved by toggling the interaction frame of the average Hamiltonian between x-, y-, and z-directions [67]. As long as MAS is slow compared to the toggling of the interaction frame, MAS and multi-pulse irradiation can be combined for averaging of anisotropic interactions (CRAMPS) [68]. For faster MAS rates above 30 kHz, where sample spinning may interfere with CRAMPS, homonuclear decoupling can also be achieved by applying phase-modulated spin lock fields alternating with time windows during which data points are acquired [69, 70].

For heteronuclear decoupling, the most straightforward decoupling technique is on-resonance rf irradiation of protons with an rf field of constant amplitude and phase (continuous wave decoupling, CW) [71] during evolution or detection periods. As CW decoupling leads only to the removal of zero order terms, better decoupling efficiencies can be obtained by phase modulations in the rf field [72–74]. However, as rf irradiation synchronized with the MAS rate may lead to interference of motion in real and spin space and thus to a recoupling of unwanted dipolar interactions instead of a decoupling, at high MAS rates extreme care has to be taken that the decoupling power is not in the same order of magnitude as the MAS rate. At ultrafast MAS rates it may be convenient to choose a decoupling power which is much lower than the MAS rate [62, 75].

4.2 Recoupling

Magic angle spinning narrows spectral lines and thus enhances resolution. On the other hand, valuable information about internuclear distances – encoded in anisotropic couplings – is lost. It is highly desirable to retrieve these interactions even in the presence of MAS for defined time intervals. Anisotropic interactions like homo- or heteronuclear dipolar couplings can be reintroduced selectively utilizing a second periodic modulation of the respective Hamiltonian which interferes with the averaging by MAS. This procedure is called recoupling (for recent reviews see [76, 77]). Homonuclear and heteronuclear dipolar recoupling can easily be incorporated into 2D experiments. Cross peaks between different spins are an indication of a dipolar coupling, i.e., spatial vicinity between nuclear spins.

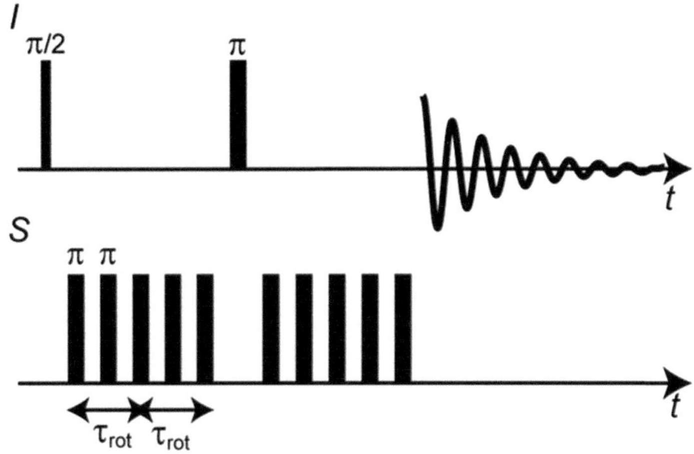

Fig. 3 In REDOR-recoupling, two phase-alternating 180°rf-pulses every rotor period τ_{rot} on the heteronuclear *S*-channel average out the effect of magic angle spinning on heteronuclear dipolar couplings. The 180°-pulse in the middle of the recoupling block on the *I*-channel refocuses the chemical shift Hamiltonian

One of the most illustrative examples for recoupling of heteronuclear dipolar couplings is rotational echo double resonance (REDOR) [78]. In this recoupling scheme, the dipolar Hamiltonian is inverted twice per rotor period by applying 180°-pulses on one of the two rf channels (Fig. 3). This periodic inversion interferes with the periodic averaging of the dipolar coupling by MAS. As a consequence, a scaled dipolar Hamiltonian is retained for each spin pair which depends on the orientation of the two spins within the rotor. Whereas the classical REDOR-recoupling results in a distance-dependent dephasing of dipolarly coupled spins, rotational echo double resonance recoupling can also be utilized for coherence transfer (TEDOR) [79]. Recent improvements of this technique allow for selective recoupling of defined spin pairs circumventing dipolar truncation by means of selective pulses or z-filtering [80].

The technique applied most frequently for heteronuclear dipolar recoupling is cross polarization. Heteronuclear dipolar interactions can facilitate polarization transfer between spins in the doubly rotating frame if two rf spinlock fields with appropriate field strengths are applied simultaneously on both channels. The ratio between the rf field amplitudes ${B_1}^{H}$ and ${B_1}^{X}$ on the proton and the heteronuclear rf channel under MAS at a spinning frequency ν_{rot} has to match the Hartmann–Hahn condition:

$$B_1^{H} \cdot \gamma_H = B_1^{X} \cdot \gamma_X \pm n \cdot \nu_{rot}$$

The polarization transfer from protons to low-gamma nuclei leads to a polarization enhancement of γ^{H}/γ^{X}. Therefore, in almost all experiments, cross polarization from protons to low-gamma nuclei is used for excitation instead of a 90° pulse on the low-gamma nucleus itself. Further, long T_1 relaxation times of low-gamma nuclei, may be circumvented by initial cross polarization from protons.

Cross polarization can also be used for magnetization transfer between low-gamma nuclei like ^{15}N and ^{13}C as a mixing sequence in a multidimensional NMR experiment. Due to the substantial shift separation of about 120 ppm between carbonyl and Cα carbon atoms in proteins, the CP transfer can be tailored specifically to NCα- or NCO-transfer (SPECIFIC-CP) [81]. In protein NMR spectroscopy such a heteronuclear correlation experiment may be incorporated into multidimensional experiments and thus be used to residue-specific resonance assignment in NMR spectroscopy of immobilized peptides or proteins [82].

Likewise, homonuclear couplings can be reintroduced by interfering with the magic angle spinning: the simplest recoupling scheme is rotational resonance recoupling which occurs if the rotor frequency is equal to the chemical shift difference of two dipolarly coupled spins, a phenomenon which was discovered as early as 1966 by Andrew [83] and subsequently explained and exploited for distance measurements in the solid state by Levitt, Griffin and coworkers [84].

Applying a spin-lock pulse with an rf amplitude equal to one or two times the rotor frequency leads to a recoupling of heteronuclear dipolar interactions (rotary resonance recoupling) [85]; a spin-lock pulse with an amplitude half of the rotor frequency recouples homonuclear dipolar couplings (homonuclear rotary resonance HORROR) [86].

Homonuclear dipolar coupling can also be recoupled by a variety of rotor-synchronized symmetry-adapted pulse schemes developed by Levitt [87]. In these sequences, the rf amplitude is equal to a fixed ratio of the rotor frequency, and the phase of the rf field is incremented or alternated at a defined rate by phase increments specified by the symmetry of the recoupling scheme.

Broadband homonuclear recoupling schemes which lead in zeroth order approximation to a dipolar Hamiltonian containing single- or double-quantum two-spin operators, however, suffer from dipolar truncation [88–90]: in multi-spin systems, small dipolar couplings between distant spin pairs are attenuated by a strong dipolar coupling to a third (closer) spin. Thus, those recoupling sequences are not suitable for obtaining multiple long-range distance constraints in extensively isotope labeled biomolecules. One remedy against dipolar truncation is the band-selective recoupling of dipolar couplings, possibly in combination with sparse isotope labeling [91]. Alternatively, second-order recoupling sequences like proton-driven spin diffusion and variants thereof as well as proton-assisted second-order recoupling techniques described in the following paragraph have been discovered and developed for obtaining structural information in biomolecular solid-state NMR spectroscopy.

4.3 (Proton-Driven) Spin Diffusion and Second-Order Recoupling for Polarization Transfer

Spectral spin diffusion, i.e., the exchange of longitudinal magnetization between dipolarly coupled nuclear spins of roughly the same energy via flip-flop transitions, can also be utilized for magnetization transfer in homonuclear correlation experiments. Although the homonuclear dipolar Hamiltonian for an isolated spin pair is averaged to zero under MAS, dipolar couplings in a multi-spin system lead to non-zero higher order terms in the average Hamiltonian. Thus, spin diffusion between protons can be used for magnetization transfer in homonuclear 2D ^{1}H-^{1}H correlation experiments [92–94] or be detected indirectly on heteronuclei [95, 96]. Likewise, dipolar couplings between ^{13}C or ^{15}N spin pairs have non-zero higher order terms when coupled to an abundant proton spin network. As a consequence, proton-driven spin diffusion (PDSD) is affected by dipolar truncation to a much lesser degree than active zero-order recoupling schemes [97]. Thus, in principle, structure determination based solely on distance constraints obtained by PDSD is possible [98, 99]. In order to facilitate proton-driven spin diffusion at high MAS spinning speeds, heteronuclear dipolar couplings to protons can be actively recoupled, as exploited in the dipolar assisted rotational resonance (DARR) mixing sequence [100–102]. For fast MAS rates, heteronuclear recoupling to protons can be made more robust by varying the phase angle of proton irradiation using the phase-alternated recoupling irradiation scheme (PARIS) [103].

At fast MAS rates, the energy mismatch of a zero-quantum transition due to a large chemical shift difference between both carbon nuclei can be compensated for by irradiating the proton spins with an rf field strength equal to the sum of the spinning speed and the chemical shift difference. This mixed rotational and rotary resonance recoupling was termed MIRROR spin diffusion [104]. Finally, the proton-assisted recoupling (PAR) [105] scheme, which involves irradiation on both proton and carbon channels avoiding all rotary resonance and Hartmann-Hahn conditions, is characterized by a three-spin effective Hamiltonian and is thus also relatively robust with respect to dipolar truncation. A heteronuclear three spin order variant, which involves three different types of nuclei by irradiating three rf channels, is called proton-assisted insensitive nuclei cross polarization (PAIN-CP) [106]. This recoupling scheme may lead to valuable long-range N–C distance information in extensively isotope labeled samples.

4.4 Dynamics

Solid-state NMR spectroscopy techniques traditionally rely on dipolar couplings between nuclei for coherence transfer. However, proteins in lipid bilayers or in proteinaceous aggregates may exhibit residual large-scale or local motions which may lead to a modulation of the anisotropic interactions on the respective time scale: submicrosecond motions lead to an averaging of the dipolar couplings and thus to a line-narrowing, whereas motions in the millisecond regime may be detected in 2D exchange experiments. For motions between these time scales, the time-dependent modulation interferes with the NMR time scale and thus leads to a line-broadening which is not averaged out by MAS. Highly flexible protein regions may selectively be studied by classical solution NMR techniques where the INEPT (Insensitive nuclei enhanced by polarization transfer) [107] based coherence transfer without proton decoupling leads to selective excitation of mobile regions [108–111]. Anisotropic motions such as sidechain rotations as manifested in the order parameter, can be obtained from recoupling of dipolar interactions and chemical shifts (DIPSHIFT) [112].

On the other hand, exchange broadening by motions in an intermediate dynamic regime may lead to complete disappearance of signals, which may become visible at lower temperatures [113, 114].

5 Applications

5.1 Amyloid Fibrils

A field of research, on which solid-state NMR spectroscopy had a tremendous impact during the past decade, is the structural study of amyloid fibrils. Solid-state NMR spectroscopy can provide information on several aspects of the cross-β core structure of amyloid fibrils such as the localization of β-strands within the amino acid sequence and their relative arrangement within protofilaments and at the protofilament interface. Even high-resolution structures for the fibril core have been determined. An overview over emerging central motifs for amyloid structure is given in Fig. 4.

Solid-state NMR spectroscopy can furthermore be employed to identify alternative non-cross-β core structures and to characterize the fibril periphery. It also reports on dynamic processes in amyloid fibrils permitting, e.g. the differentiation between segments of static and dynamic disorder.

In the following we give a brief overview of selected amyloid systems, whose structural characterization has greatly benefited from the use of solid-state NMR spectroscopy. For a more detailed insight we refer to recent review articles [3, 36, 37, 115].

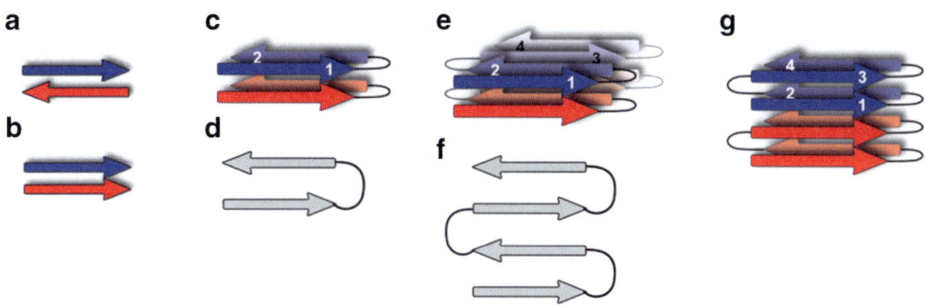

Fig. 4 Structural motifs underlying amyloid formation: (**a**) antiparallel β-sheet, (**b**) parallel β-sheet, (**c**, **d**) parallel β-sandwich consisting of two β-sheets connected by a 180° turn as suggested for Aβ(1–40), (**e**, **f**) superpleated parallel cross-β-structure with more than two β-sheets as suggested for Ure2p and Sup35, (**g**) parallel β-helix consisting of two β-sheets with one monomer involved in two layers. In case of *gray arrows*, a cross section of the fibril axis is shown, whereas *colored arrows* indicate a side view parallel to the fibril axis. Every monomer is displayed in a *different color*

5.1.1 Aβ Peptide

The β-amyloid peptides Aβ(1–40) and Aβ(1–42) are the major constituents of senile plaques and cerebrovascular amyloid deposits found in the brains of patients with Alzheimer's disease (AD) and cerebral amyloid angiopathy [116]. Aβ is an alternative cleavage product of the amyloid precursor protein (APP), a membrane protein of still unknown function [117]. Fibrils of Aβ and smaller peptides thereof have been extensively studied by solid-state NMR spectroscopy. Although a β-sandwich always seems to be the underlying motif, fibril structure details are not only highly susceptible to polymorphism but also depend critically on preparation conditions. Tycko and coworkers identified two different polymorphic forms of Aβ(1–40) fibrils, characterized by different chemical shifts and different fibril morphologies. These two conformations are obtained under different fibrillation conditions and can be transferred to the next fibril generation by seeding [42]. Two different supramolecular arrangements were suggested for these polymorphs [43].

A distinct type of polymorphism was observed in the Aβ(1-40)-Iowa mutant D23N [44, 118]. In the same fibril preparation both parallel and antiparallel β-structures coexisted, the latter representing the major conformation. This finding illustrates that a single disease-related mutation can have extensive consequences for amyloid structure.

As an approach to study the structure of Alzheimer's disease-related Aβ fibrils, Aβ(1-40) fibril growth was directed by seeding with fibrils extracted from brain tissue of deceased AD patients. The resulting NMR data supported a β-strand-turn-β-strand motif similar to *de novo*-generated Aβ-fibrils. The chemical shifts, however, were significantly different from those observed for unseeded fibrils, suggesting structural differences between brain-derived and synthetic Aβ fibrils [45]. Recently, Reif et al. reported another fibril type composed of asymmetric dimers [119], whereas Bertini,

Mao et al. further extended the structural diversity in the U-shaped β-strand-turn-β-strand motif of Aβ-fibrils, pointing towards a complex picture of Aβ-fibrillation [120].

Solid-state NMR studies of Aβ oligomers or protofibrils are extremely challenging due to their metastability. However, NMR data obtained for Aβ(1–40) [121] and Aβ(1–42) [122] oligomers demonstrate chemical shifts similar to the fibrillar peptides. Aβ(1–40) protofibrils could be stabilized by an antibody-derived fusion protein [123, 124]. The corresponding protofibril chemical shifts were indicative of a β-sheet structure in the same sequence region as for mature Aβ fibrils. However, the protofibril β-strands were less extended, and the analysis of chemical shifts suggested a closer relation to oligomers than to mature fibrils.

5.1.2 α-Synuclein

Amyloid fibrils of α-synuclein are the main component of Lewy bodies which are the pathological hallmark of Parkinson's disease [125]. The physiological function of the 140-residue α-synuclein is still elusive. Three point mutations associated with early onset familial Parkinson's disease have been identified so far [126–128]. Fibrillar α-synuclein was the first amyloid from a full-length protein studied by solid-state NMR spectroscopy [110, 129] and the spectral quality has subsequently encouraged further studies [130–134]. The protein is capable of forming a high variety of polymorphic forms. In different studies at least four different fibril types have been obtained even under rather similar fibrillation conditions. These deviating fibril types are characterized by differences in chemical shifts of identical residues as well as the exact location of well-ordered β-strands and of statically disordered parts in the rigid *N*-terminus. Unambiguous assignments for the first 30 *N*-terminal residues could only be obtained in one case [134]. In contrast, the mostly negatively charged 40 C-terminal residues were always found to be flexible, lacking a defined secondary structure. Analyzing NC-transfer or CC-transfer spectra of differentially isotope labeled fibril units, residues 40–90 were identified to form parallel in register β-sheets [3, 133] which are in agreement with EPR-studies [135].

The fibrillation kinetics of an A30P α-synuclein mutant was observed to be substantially slower, although chemical shifts were identical to one wild-type fibril form grown under identical conditions [136]. Studies on A53T α-synuclein fibrils revealed an extended β-sheet core [113] and slight perturbations of the chemical shifts, whereas for the E46K-mutation larger chemical shift deviations were observed [137].

Two artificial supertoxic mutants with one or three β-sheet breaking proline substituents, respectively, were designed based on the location of the β-sheets identified by solid-state NMR spectroscopy [32]. As expected, *in vitro* fibrillation as well as aggregation in HEK cells was retarded in both mutants, especially in the triple mutant, shifting the equilibrium towards oligomeric intermediates. Corresponding solid-state NMR spectra revealed a decreased rigid β-sheet core for the single mutant. For the triple mutant, a high degree of disorder was determined as indicated by broad and featureless lines with secondary chemical shifts characteristic for β-sheets. A higher toxicity of these artificial mutants was confirmed by expression in *Drosophila melanogaster*, *Caenorhabditis elegans*, and cultured mammalian neurons.

5.1.3 Yeast Prions with Glutamine/Asparagine-Rich Prion Domains

Certain proteins in yeast cells or fungi can adopt alternative β-sheet-rich conformations leading to epigenetic variations of protein function and thus to different phenotypes [138]. Since the underlying metastable conformational changes can be transferred horizontally, these proteins are termed yeast prions. *Saccharomyces cerevisiae* prions such as Ure2p, Sup35, Rnq1p, and Mod5 consist of a functional globular domain and a prion domain which is rich in asparagine and glutamine residues [139, 140]. In the prion conformation, the prion domains adopt a parallel, in-register β-sheet structure [46, 141–144] stabilized by Q/N-polar zippers [145]. For Sup35p and Ure2p, the ability to form prions was demonstrated to be retained after shuffling of the prion domain sequence, probably due to the high content of Q/N-residues allowing for different patterns of forming parallel, in-register β-sheet structures [146, 147].

Yeast prions show a high degree of structural polymorphism which may be linked to the existence of different yeast prion strains. Polymorphism has been detected by solid-state NMR spectroscopy for full-length Sup35p as well as for a heptapeptide from Sup35p [144, 148, 149]. Yeast prion-seeded samples of Ure2p exhibited different solid-state NMR spectra than non-seeded samples [46]. Likewise, NMR-signals from the isolated Ure2p-prion domain differed from the full-length protein pointing to extensive interactions of the globular C-terminal domain with the prion domain [46, 150].

5.1.4 Functional Amyloid: The Yeast Prion HET-s

The prion protein HET-s from *Podospora anserina* is a functional prion involved in heterokaryon incompatibility and, so far, the only yeast prion protein for which the prion state leads to a gain of function instead of a loss of function. Solid-state NMR spectra of HET-s exhibit remarkably high resolution indicative of high local order and structural homogeneity. This permitted the determination of a high-resolution structure [151–153]. In contrast to all other amyloid fibrils studied so far, the HET-s prion domain forms a β-solenoid with two windings per monomer and a triangular hydrophobic core (Fig. 5). Studies on the full-length protein revealed that the globular domain is not well structured but can be considered as a molten globule [154].

Recently, HET-s in the amyloid form was used as model fibrils for the study of Congo Red binding. The binding interface was determined by cross polarization from unlabeled Congo Red to fully deuterated HET-s fibrils [155]. Based on the complex structure, a non-congophilic HET-s mutant could be designed, providing a structural

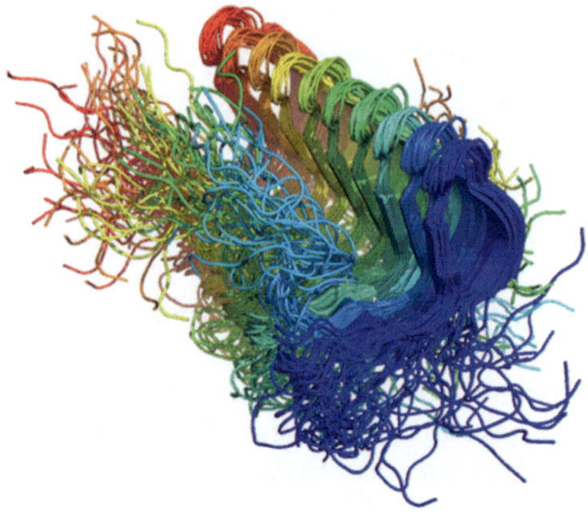

Fig. 5 Structure of HET-s (218–289) in its amyloid form as revealed by solid-state NMR spectroscopy. The depicted bundle of 20 conformers was deposited in the Protein Data Band with accession code 2RNM [152]

rationale for the different sensitivity of amyloid-binding dyes for amyloids of different proteins or different prion strains/variants.

5.2 Membrane Proteins

A second field in structural biology, where solid-state NMR has proven to be useful, is the study of membrane proteins. As lipid bilayers – the natural environment of membrane proteins – tend to form liposomes or vesicles with rotational correlation times too large for solution-state NMR spectroscopy, solid-state NMR spectroscopy has been demonstrated to be a viable method to gain insight into complex membrane protein systems. In the following section we give a brief overview over selected systems where solid-state NMR could provide useful information. For a more comprehensive overview we refer to recent review articles [4, 146, 156, 157].

5.2.1 Influenza Ion Channel M2

The 97-residue M2 protein of the influenza A virus forms a tetrameric proton channel which is activated at low pH values. It has been shown to play an important role in virus replication. After infection of the host cell by endocytosis, the low endosomal pH opens the channel. The subsequent acidification of the viral core triggers the dissociation of the virus and thus induces expression of the viral genes. This ion channel is a target for the antiviral drugs amantadine and rimantidine which inhibit proton conduction and thus the unpacking of the virus. The S31N point mutation in the most recent circulating virus strains results in resistance against these drugs.

Consequently, structural information about the drug binding site is of prevalent interest.

As the single transmembrane helix of the M2 protein alone, M2(22–46), is capable of tetramerization and proton conduction, it constitutes an ideal target for structural studies. Two high-resolution structures of a detergent-stabilized transmembrane helix construct bound to amantadine and of a construct containing the transmembrane helix and an adjacent amphipathic helix bound to rimantidine have been determined by X-ray crystallography [147] and solution NMR spectroscopy [158], respectively. However, both studies revealed contradicting drug binding sites, indicating that structure and function are sensitive to the lipid environment. An intense solid state NMR spectroscopy investigation of a 25-residue construct of the transmembrane helix reconstituted into liposomes solved the question. Solid-phase synthesis of the short peptide construct allowed for site selective isotope labeling of residues of interest and thus accurate resonance assignments in spite of line widths larger than 1 ppm. Chemical shift mapping of selected residues upon amantadine binding revealed only slight chemical shift deviations for most residues of the channel [159] except for Ser31 [160] (Fig. 6). Subsequent ^{13}C-^{2}H REDOR measurements on a site-selectively ^{13}C-labeled M2 transmembrane helix complexed with fully deuterated amantadine revealed one strong binding site at Ser31 and one weak binding site at the C-terminal membrane side [161]. The orientation of the drug could be inferred from ^{2}H-lineshapes. Finally, studies on a larger construct, including the amphipathic helix, revealed that the strong binding site persists in the longer construct while the weak binding site is obstructed by the amphipathic helix in liposome samples [162]. A comparison of spectra from full-length M2 in native *E. coli* membranes, liposome-reconstituted full-length M2, and a liposome-reconstituted construct containing the transmembrane and the amphipathic helices yielded comparable chemical shifts for the three complexes, thus confirming structural similarity of the shorter constructs to the full-length protein [163].

Other questions of interest comprise the mechanism of proton conduction, selectivity, and pH gating. One single His37 residue in the transmembrane helix is responsible for proton selective conduction at low pH values, whereas at high pH values four His side chains constrict the channel. Two models for proton conduction had been proposed. The so-called shutter model suggests that, at low pH, protonation of the His residues leads to pore opening by electrostatic repulsion of the four positively charged rings and formation of a water wire which conducts protons via the Grotthus mechanism. According to the shuttle model, His 37 actively shuttles the protons into the virion by subsequent protonation and deprotonation accompanied by conformational ring flip motions (see e.g. ref [164] and references therein for more details).

In a recent investigation, Hong and coworkers studied the transmembrane domain of M2 in liposomes at high and low pH [164]. At high pH, distinct resonances for two different tautomers of the neutral imidazole ring were resolved, and cross peaks

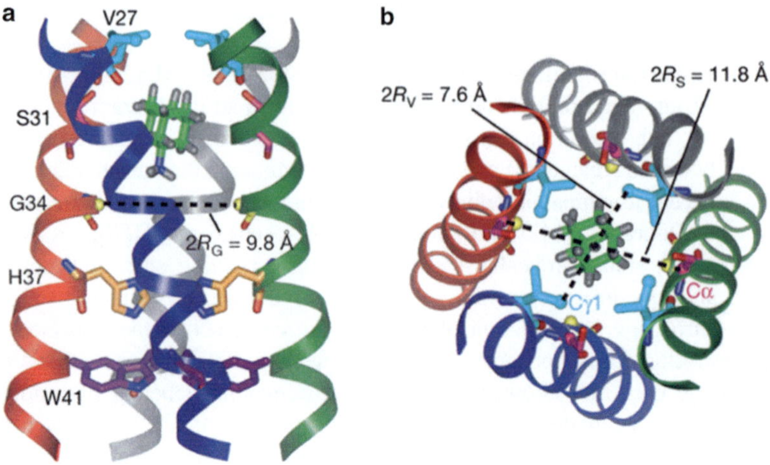

Fig. 6 (**a**) Side view showing the transmembrane part of tetrameric M2 (22–46) with amantadine (Amt) bound to the high-affinity luminal site, (**b**) top view showing the Ser 31 and Val 27 pore radii of the transmembrane part of M4 binding AMT. The orientation of Amt is slightly tilted from the channel axis. The time-averaged Amt orientation is parallel to the channel axis. Reprinted with permission from Macmillan Publishers Ltd: Nature [161], copyright (2012)

between those resonances in the PDSD spectra indicated an edge-to-face π-stack of the four histidines which obstructs the channel. At low pH the protonated imidazolium rings were found to undergo microsecond ring flip reorientations (faster than the NMR time scale), as measured by the order parameters obtained from recoupled one-bond N–C and H–C dipolar couplings of the aromatic rings, thus confirming the shuttle model. Determination of HN and HC dipolar couplings at low temperature, i.e., when the reorientation of the rings was frozen revealed slightly enlarged bond lengths, indicative of hydrogen bond formation. Proton spin diffusion experiments identified water as the sole hydrogen bonding partner [165]. Finally, different polarization transfer rates from water to the protein in the open state, at low pH, in the closed state at high pH, as well as in the amantadine-blocked state revealed a larger water contact surface for the open channel and a reduced water contact surface for the blocked channel [166].

A recent study of a construct containing the transmembrane helix and the amphipathic helix of the S31N mutant [167], which was reversely isotope labeled for four hydrophobic residues, facilitated sequential resonance assignments for 23 residues. All secondary chemical shifts were found to be characteristic for α-helical secondary structure. Peak doubling of most resonances as well as cross peaks between resonance pairs in PDSD spectra indicates that the tetramer is a dimer of asymmetric dimers instead of a symmetric homotetramer. Upon addition of amantadine, no significant shift changes were observed, which is in agreement with the drugs resistance of this mutant.

5.2.2 Seven Transmembrane Helix Receptors

Seven transmembrane helices (7-TM) are a structural motif common to a large family of photo- and chemoreceptor proteins. Most prominent examples of 7-TM proteins include (bacterial) rhodopsins and G protein-coupled receptors (GPCRs).

5.2.2.1 Bacterial Rhodopsins

In archaea and bacteria a family of 7-TM proteins with retinal as a prosthetic group serves two distinct functions. First, a light-driven energy conversion, such as carried out by bacteriorhodopsin from the extreme halophile *Halobacterium salinarum* [168] or by proteorhodopsin [169], converts light energy into an ion gradient, building up an electric potential across the membrane. Second, phototactic light sensing, such as performed by sensory rhodopsin [170], activates a signal transduction chain via a bound transducer molecule.

Rhodopsins of both classes have been successfully investigated using NMR spectroscopy. While progress in solution-state NMR has enabled the *de novo* structure determination of the two 7-TM proteins sensory rhodopsin II [171] and proteorhodopsin [172] solubilized in their monomeric form in detergent micelles, the power of solid-state NMR lies in the investigation of 7-TM receptors in their native environment and oligomeric state. In this respect the sensory rhodopsin system offers a remarkable example for the potential of solid-state NMR. In its native oligomeric state, the 7-TM receptor sensory rhodopsin, upon light excitation triggers a signal transduction chain via a tightly bound transducer protein, which is homologous to the two-component system of eubacterial chemotaxis. So far, the native oligomeric state of the receptor and transducer complex has not been formed in detergent micelles [173], impeding the use of solution-state NMR. In addition, pronounced intermolecular crystal packing contacts are found in the X-ray structure in close proximity to the receptor-transduced binding interface [174]. Solid-state NMR could be used to characterize the heterodimeric receptor–transducer complex in a lipid bilayer setting [175]. In this native-like condition, solid-state NMR data identified a considerably larger binding interface than found in the crystal structure, contributing to a better understanding of phototactic signal transduction.

Further examples for the use of solid-state NMR to study 7-TM proteins include the light-activated proton pump bacteriorhodopsin which could be studied in detail in its native environment, the purple membrane. Valuable high-resolution insights into the structure of the functional core – in particular the configuration of the covalently bound retinal at different states in the photocycle – could be obtained [176–180].

More recently several other bacterial rhodopsins such as proteorhodopsin and Anabaena sensory rhodopsin (ASR) have been investigated with solid-state NMR. Remarkable spectral resolution and sensitivity resulting in nearly complete resonance

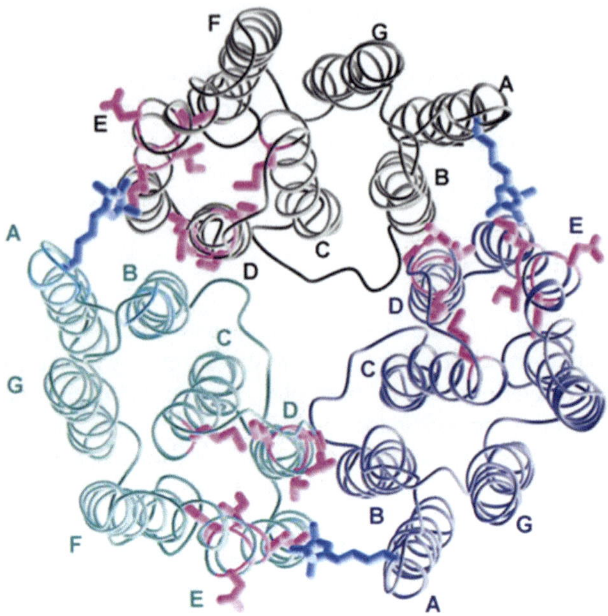

Fig. 7 Trimer model of S26CR1 Anabaena sensory rhodopsin (ASR). The individual monomers are shown in different colors. R1 side chains are shown in *blue*. Side chains of residues experiencing large intermolecular paramagnetic relaxation enhancement (PRE) effects and therefore spatially close (<15 Å) to the nitroxide of a neighboring monomer are shown in *magenta*. Adapted with permission from [189]. Copyright (2012) American Chemical Society

assignments were obtained [181–184]. Based on these data, specific insights into, e.g., the effects of hydration water on molecular dynamics [185] or the structural properties of the inner core region of these 7-TM systems could be generated [186–188]. In addition, protein–protein interactions could be studied with paramagnetic relaxation measurements. Using this approach, reliable information about the oligomeric state as well as the quaternary structure of ASR in a lipid bilayer environment was obtained (Fig. 7) [189].

Overall, these exciting results indicate that solid-state NMR spectroscopy will provide increasingly detailed information about structure, dynamics, and function of rhodopsins in a near-native setting within the next few years.

5.2.2.2 G protein-Coupled Receptors

Acting as a central interface between external stimulus and cellular response, G protein-coupled receptors (GPCRs) control the majority of intercellular signaling cascades in humans. GPCRs play an important role not only in a large number of signaling cascades but also in very diverse physiological processes such as vision and smell as well as the regulation of blood pressure, body weight, and cell death. Due to their essential function in human physiology, it is not surprising that between 30% and 50% of all modern drugs are estimated to target GPCRs [190–192]. Although sharing the same topology of seven transmembrane helices with the class of (bacterial) rhodopsins, the investigation of GPCR structure is considerably more challenging, predominantly due to difficulties in sample preparation.

However, tremendous efforts have been carried out to solve the 3D-structure of GPCRs. Recently a breakthrough in protein crystallization attempts led to publication of a growing number of GPCR crystal structures (see, e.g., [193–195] for recent reviews). To stabilize crystal growth, artificial cofactors and protein engineering, such as large modification of potentially important loop segments, turned out to be necessary. Hence, additional information obtained under more native-like conditions is very desirable to provide complimentary insights into receptor dynamics as well as ligand binding and G-protein interactions [196].

While a number of NMR studies of GPCRs and their ligands have been carried out (reviewed, e.g., in [146] and [197]), no well-resolved NMR spectrum of a solubilized GPCR could be obtained so far, which is related to low protein expression yields, high amounts of misfolded receptor, and lack of stability when incorporated into a non-native membrane mimetic. On the contrary, solid-state NMR was used to study functional GPCRs with high precision [198–202]. In addition to the use of a more stable lipid bilayer environment, the advantage of solid-state NMR for the investigation of GPCRs is the ability to study non-deuterated membrane proteins. This enables the use of eukaryotic expression systems such as human or insect cell lines which do not tolerate a high level of deuterated buffer. While relaxation effects impede high resolution solution-state NMR studies of solubilized GPCRs without a significant degree of deuterium labeling, solid-state NMR does not face this problem and hence allows the characterization of functional GPCRs expressed in eukaryotic systems. In particular, in combination with the structures obtained using X-ray crystallography, it can be anticipated that solid-state NMR will provide critical information about ligand sensing and signal transduction in GPCRs.

6 Sensitivity Enhancement By Dynamic Nuclear Polarization

6.1 Theoretical Background

The fact that electron spins have a gyromagnetic ratio about 660 times higher than proton spins is exploited in dynamic nuclear polarization (DNP).

As already proposed by Albert Overhauser [203] in 1953 and subsequently confirmed experimentally by Carver and Slichter [204, 205], electron spin polarization in solid metals can be transferred to nuclear spins if the unpaired conducting electrons are saturated by irradiation with the corresponding electron paramagnetic resonance (EPR) frequency. As a result, nuclear spin polarization may be enhanced by the factor γ_e/γ_n. Likewise, stable organic radicals can be used as polarizing agents.

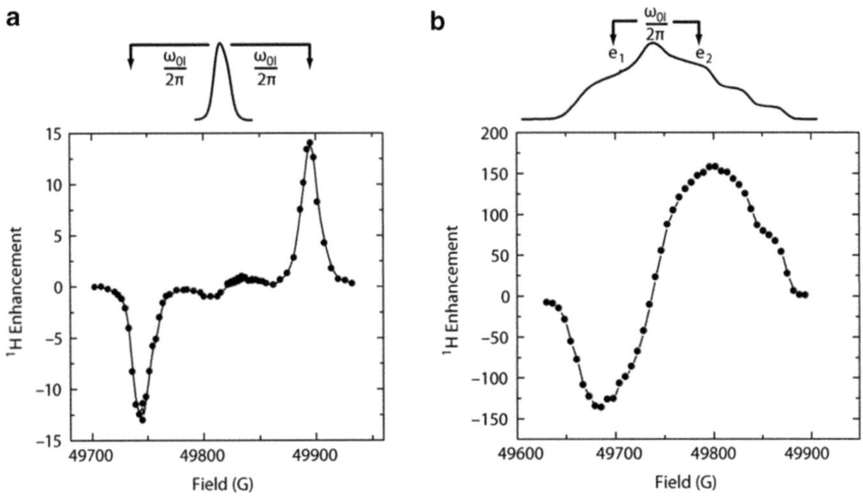

Fig. 8 Experimental ^{1}H DNP enhancement profiles for the SE and the CE mechanisms showing the positions of positive and negative enhancement and their dependence on the microwave irradiation frequency (or, rather the magnetic field for a given microwave frequency) as well as on the electron and nuclear Larmor frequencies ω_{0S} and ω_{0I}, respectively. (**a**) A typical SE enhancement profile obtained with 40 mM trityl. (**b**) A typical CE enhancement profile obtained with 10 mM TOTAPOL (20 mM electrons). The EPR spectrum of each radical is shown on top. The *lines* connecting the data points are to guide the eye. Reprinted with permission from [206]. Copyright 2012, American Institute of Physics

Depending on the EPR-line width and the strength of the external magnetic field (and thus the line width of the EPR transition) the polarization transfer between electrons and nuclear spins in solid samples can occur via the Solid-Effect (SE), the Cross-Effect (CE) or Thermal Mixing (TM).

SE is a two-spin process, which relies on the excitation of forbidden zero or double quantum electron-nuclear two-spin transitions. As these transitions are forbidden, their excitation requires high power microwave irradiation. The efficiency of the transfer scales with the inverse square of the nuclear Larmor frequency. Another prerequisite for a selective irradiation of the zero or double quantum transition using SE DNP is a line width of the corresponding EPR-transition smaller than twice the nuclear Larmor frequency (Fig. 8). Nevertheless, by enhancing the microwave power with a resonator, SE has been demonstrated to lead to a sensitivity gain of about factor 128 at a field of 5 T [207]. TM and CE constitute three-spin transitions involving two electron spins and one nuclear spin at high radical concentrations. They are effective if the line width of the homogeneously or heterogeneously broadened EPR-transition is larger than the nuclear Larmor frequency, such that a three-spin transition is energetically neutral (similar to proton spin diffusion) [208, 209]. While thermal mixing is most effective at low fields, where the field-dependent line broadening due to the g-factor anisotropy is small, the cross effect can

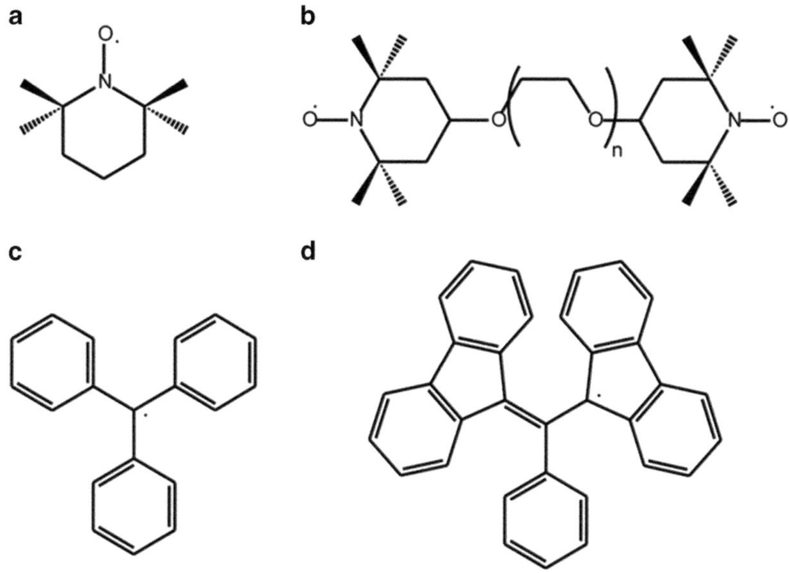

Fig. 9 Structural formula of the radicals (**a**) TEMPO, (**b**) TOTAPOL, (**c**) trityl, (**d**) 1,3-bisdiphenylene-2-phenylallyl-radical (BDPA)

facilitate efficient polarization transfer at high magnetic fields and low temperatures, especially if biradicals are used.

The recent success of DNP as a hyperpolarization technique in solid-state NMR spectroscopy is mainly due to the following developments [210].

For high-resolution NMR investigations high magnetic fields are needed. Consequently, high-power high-frequency microwave sources are necessary to saturate the broad EPR lines at these fields. This can be facilitated by gyrotrons, which are able to produce stable microwaves of frequencies of up to 263 GHz when operating at the fundamental frequencies. Higher microwave frequencies can be obtained by operating a gyrotron with a given magnetic field at the second harmonic. This results in frequency doubling and facilitates microwave frequencies of up to 526 GHz, corresponding to a magnetic field strength of 18.8 T or a proton Larmor frequency of 800 MHz [211–213]. As the spin lattice relaxation times of nuclear spins have to be sufficiently long to facilitate polarization transfer, low temperatures below 90 K are required. This condition can now also be met for applications with MAS.

Paramagnetic centers used as source for the polarization transfer need to be compatible with the polarization mechanism. For SE-DNP, which is effective for isolated spin centers, a polarizing agent characterized by a narrow EPR transition line is needed. For this purpose, derivatives of the trityl radical or 1,3-bisdiphenylene-2-phenylallyl (BDPA) (Fig. 9) have been demonstrated to be well suited [214]. If polarization is to be provided by external paramagnetic dopants, a three-spin transition by the CE transfer mechanism is the method of choice. Then at least two paramagnetic centers have to be strongly dipolarly coupled. This prerequisite is fulfilled by stable biradicals such as TOTAPOL [215–217], which consists of two closely linked TEMPO-radical units.

For recent reviews of DNP in solids at high fields we refer to the following review articles [210, 218]. In the following section we will give an overview about recent developments and applications of this emerging technique.

6.2 Applications of DNP to Biomolecular NMR Spectroscopy

The applicability of DNP to biomolecules in the solid state was first demonstrated using lysozyme in a frozen glycerol/water matrix doped with TEMPO monoradicals. In this study, helium gas cooled to 40 K served as bearing and drive gas for MAS [219]. Subsequently, hyperpolarization by DNP has successfully been applied to a virus particle [220], to multidimensional heteronuclear spectroscopy of protein microcrystals of amyloidogenic peptides [221], and to a membrane receptor trapped in different stages of the photocycle [176, 177, 222].

Intending to improve sample preparation and to explore different experimental conditions, it was then realized that one potential limitation to biological application may be the line width. Since low temperatures below 90 K are required for efficient polarization transfer, lines may be severely broadened, either homogeneously or inhomogeneously:

At low temperatures, especially mobile side chains in a protein can be frozen in multiple static conformations, resulting in inhomogeneous line broadening. Particularly in spectra of hydrated microcrystals, line broadenings of up to 3–4 ppm are observed upon cooling to below a temperature of 200 K [59, 60]. For extensively deuterated samples, however, DNP enhancements of one order of magnitude can still be obtained at temperatures around 180 K, maintaining a reasonable resolution of 2D-spectra which is significantly increased compared to 90 K spectra [223].

In addition to inhomogeneous line broadening, the presence of paramagnetic dopants may lead to shortened T_2 nuclear relaxation times and thus a homogeneous line broadening. When the polarizing agents, however, are spatially separated from the nuclei of interest, narrow lines can also be observed with DNP [224]. Thus, the spatial distribution of radicals within the sample as well as their concentration also influences the resolution of spectra [225].

Signal enhancement for more elaborate NMR sequences involving multiple CP-transfer steps depends critically on the nuclear relaxation rates in the rotating frame of all nuclei involved. As these relaxation rates can be drastically reduced in the presence of paramagnetic dopants, it is not surprising that signal enhancement does not increase monotonically with increasing radical concentrations, but reaches a maximum at rather moderate radical concentrations [225].

Initial DNP NMR experiments on a complex biochemical system, i.e. amyloid fibrils of the model peptide GNNQQNY, yielded encouraging results [226]. Addition of biradicals to the sample did not result in significant line broadening and chemical shifts were preserved at low temperatures. Since chemical shifts before and after freezing the sample to 100 K were also identical, low temperatures do not appear to change the fibril structure at a molecular level.

Subsequently, for amyloid fibrils of the SH3 domain of the 83-residue phosphatidylinositol-3-kinase, valuable structural information could be provided by MAS-DNP-NMR [227]. For fibrils grown from a mixture of exclusively ^{15}N labeled monomers and monomers labeled by using [2-^{13}C]glycerol, intermolecular distance constraints were obtained from TEDOR experiments. High field spectra recorded at room temperature by conventional NMR spectroscopy provided only 30 intermolecular cross peaks for 83 residues, an observation which was ascribed to an interference of protein dynamics with decoupling, recoupling and cross polarization. In particular, no cross peaks could be observed for aromatic residues because of their twofold ring flips. In contrast, a DNP enhanced TEDOR spectrum recorded at a temperature of about 100 K and a field strength of 400 MHz ^{1}H frequency revealed many additional intermolecular ^{15}N–^{13}C cross peaks, especially in the aromatic region. Although these spectra suffered from a high number of cross peaks and a lower resolution, 20 additional distance constraints could be determined with certainty (Fig. 10).

A recent investigation of a full virion, the bacteriophage Pf1, also benefitted greatly from reduced mobility as well as signal enhancement by DNP [228]. Whereas only protein signals could be assigned in earlier conventional solid-state MAS NMR studies at room-temperature, in DNP-enhanced spectra DNA signals were also assignable. The chemical shifts of the desoxyribose ^{13}C atoms were indicative of 2′-endo/gauche conformations and anti-glycosidic bond conformations, while the chemical shifts of the DNA bases were consistent with an unusual structure with little or no base pairing, but base stacking. Further, signal splittings suggest some minor variations in the environment due to different interactions of the two (non-paired) DNA strands of the virion with the coat protein. Selected protein/DNA contacts could be identified in the spectra.

DNP enhancement was also used to investigate 40 nmol of a 25-residue signal peptide bound to the lipid-reconstituted 600-residue protein translocation complex SecY translocon [229]. Although double quantum filtering was mandatory to suppress the large natural abundance SecY background, a decent 2D-spectrum of the peptide could be obtained within 20 h of measurement time. For three of four isotope labeled amino acids the spin system could be identified by a sequential walk. The corresponding secondary chemical shifts were indicative of an α-helical secondary structure of the peptide in its bound form.

DNP signal enhancement also facilitated the structural investigation of the Asian cobra neurotoxin II bound to the ligand-gated ion channel nicotinic acetylcholine receptor (nAChR) obtained from the electric organ of an electric ray [230, 231]. Despite the low concentration of the ^{15}N- and 1,3-^{13}C- or 2-^{13}C-glycerol-labeled toxin, a decent 2D ^{13}C-^{13}C correlation spectrum could be recorded within 14 h. Interestingly, storage of the sample at −20 °C resulted in a reduction of radicals close to the membrane surface leading to a loss of polarization enhancement and a drastic resolution increase for amino acid residues close to the membrane surface.

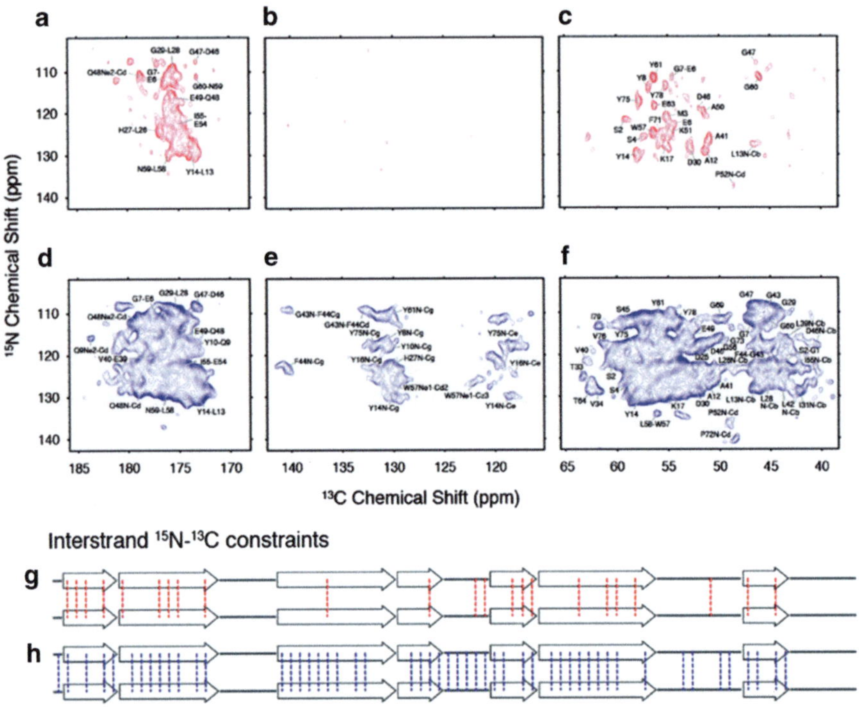

Fig. 10 Comparison between room temperature and DNP-enhanced, low temperature intermolecular correlation spectra. (**a–c**) 750 MHz intermolecular ^{15}N–^{13}C correlations in PI3-SH3 fibrils recorded at 300 K with 16 days of acquisition. (**d–f**) The identical spectral regions recorded at 100 K and 400 MHz with DNP enhancement in 32 h of signal averaging. (**g**) Illustration of the 23 interstrand contacts established from ^{13}C–^{15}N cross-peaks in the 750 MHz spectra acquired at 300 K in panels **a–c**. (**h**) The 52 interstrand contacts established from the 400 MHz DNP-enhanced spectra recorded at 100 K shown in panels (**d–f**). Reprinted with permission from [227]. Copyright (2012) American Chemical Society

The high sensitivity provided by DNP enhancement allowed for 3D-NCACX/NCOCX-based sequential walks for selectively labeled samples of the model membrane protein Mistic, even though it was studied in its native cellular membrane environment without prior purification and reconstitution. Corresponding secondary chemical shifts are in agreement with the solution-state NMR structure, indicating that the protein was well folded [232]. The integral membrane protein PagL was also studied in extracted cell walls as well as whole *E. coli* cells. A significant DNP enhancement allowed for the detection of resonances from the overexpressed PagL protein as well as from an endogenous membrane-associated lipoprotein, lipid molecules, and RNA bases, for which even inter-base pair cross correlations were observed [233].

In all previous examples, diamagnetic biomolecules were studied by solid-state NMR spectroscopy using the biradical TOTAPOL as an external polarization agent. However, about 10% of biological macromolecules possess a paramagnetic (metal)

center or a functional group or cofactor which is present as a (meta)stable radical during the reaction cycle or electron transfer process. The possibility to exploit these endogenous radicals as polarization sources appears intriguing. A first pioneering study using an endogenous radical for DNP enhancement in biomolecules involved the small electron transport protein flavodoxin. It contains a single non-covalently bound flavine mononucleotide which cycles between the oxidized quinone and the reduced radical semiquinone form. Due to the intrinsically low radical concentration, the DNP polarization transfer was governed by the solid effect, which unlike the cross effect does not rely on strong dipolar couplings between electron spins.

After deuteration of the protein to 85%, a DNP enhancement factor of 15 was achieved, demonstrating the feasibility of SE DNP using endogenous radicals despite a rather low field strength of only 212 MHz [234]. SE DNP has also been demonstrated for model complexes of the high-spin transition metals Mn^{2+} ($S = 5/2$) and Gd^{3+} ($S = 7/2$). It was found that the DNP enhancement depends strongly on the line width of the central ($-1/2 \rightarrow +1/2$) electron spin transition. For Mn^{2+}, the enhancement factor was limited to about 2 due to the hyperfine coupling to the ^{55}Mn nucleus splitting the corresponding EPR transition and thus the DNP enhancement profile into six lines. For Gd^{3+}, however, a signal enhancement factor of 13 was obtained when the line broadening of the central EPR transition by second order zero field splitting was small [235]. These studies show that utilizing metal centers as polarization source for DNP enhancement may in principle be possible.

While this review could only highlight certain examples, it can be stated that the combination of the substantial improvements, in particular in terms of the available hardware, sample preparation techniques, and novel and improved concepts of data acquisition and NMR methodology, extended considerably the limits of systems that can be studied today. For the near future, it can be anticipated that exciting new applications will emerge and that increasingly detailed insights will be obtained in more and more challenging biological systems using modern solid-state NMR spectroscopy.

References

1. Opella SJ, Marassi FM (2004) Chem Rev 104:3587
2. Andrew ER, Bradbury A, Eades RG (1958) Nature 182:1659
3. Heise H (2008) Chembiochem 9:179
4. McDermott A (2009) Annu Rev Biophys 38:385
5. Wylie BJ, Rienstra CM (2008) J Chem Phys 128:052207
6. Higman VA, Flinders J, Hiller M, Jehle S, Markovic S, Fiedler S, van Rossum BJ, Oschkinat H (2009) J Biomol NMR 44:245
7. Böckmann A (2008) Angew Chem Int Ed Engl 47:6110
8. Guo C, Geng C, Tugarinov V (2009) J Biomol NMR 44:167
9. Lundstrom P, Teilum K, Carstensen T, Bezsonova I, Wiesner S, Hansen DF, Religa TL, Akke M, Kay LE (2007) J Biomol NMR 38:199
10. Boer E, Steinborn G, Kunze G, Gellissen G (2007) Appl Microbiol Biotechnol 77:513

11. Madden DR, Safferling M (2007) Methods Mol Biol 363:39
12. Armstrong N, Sun Y, Chen GQ, Gouaux E (1998) Nature 395:913
13. Waugh DS (1996) J Biomol NMR 8:184
14. Hong M, Jakes K (1999) J Biomol NMR 14:71
15. Whittaker JW (2007) Methods Mol Biol 389:175
16. Blanco FJ, Hess S, Pannell LK, Rizzo NW, Tycko R (2001) J Mol Biol 313:845
17. LeMaster DM, Kushlan DM (1996) J Am Chem Soc 118:9255
18. Lee AL, Urbauer JL, Wand AJ (1997) J Biomol NMR 9:437
19. Kigawa T, Muto Y, Yokoyama S (1995) J Biomol NMR 6:129
20. Noren CJ, Anthonycahill SJ, Griffith MC, Schultz PG (1989) Science 244:182
21. Young TS, Schultz PG (2010) J Biol Chem 285:11039
22. Tugarinov V, Kanelis V, Kay LE (2006) Nat Protoc 1:749
23. Gardner KH, Kay LE (1998) Annu Rev Biophys Biomol Struct 27:357
24. Chevelkov V, Rehbein K, Diehl A, Reif B (2006) Angew Chem Int Ed Engl 45:3878
25. del Amo JM, Fink U, Reif B (2010) J Biomol NMR 48:203
26. Asami S, Schmieder P, Reif B (2010) J Am Chem Soc 132:15133
27. Zhou DH, Shah G, Cormos M, Mullen C, Sandoz D, Rienstra CM (2007) J Am Chem Soc 129: 11791
28. Etzkorn M, Böckmann A, Lange A, Baldus M (2004) J Am Chem Soc 126:14746
29. Mao L, Vaiphei ST, Shimazu T, Schneider WM, Tang Y, Mani R, Roth MJ, Montelione GT, Inouye M (2010) J Struct Funct Genomics 11:81
30. Havlin RH, Tycko R (2005) Proc Natl Acad Sci USA 102:3284
31. Heise H, Luca S, de Groot BL, Grubmüller H, Baldus M (2005) Biophys J 89:2113
32. Karpinar DP, Balija MB, Kugler S, Opazo F, Rezaei-Ghaleh N, Wender N, Kim HY, Taschenberger G, Falkenburger BH, Heise H, Kumar A, Riedel D, Fichtner L, Voigt A, Braus GH, Giller K, Becker S, Herzig A, Baldus M, Jackle H, Eimer S, Schulz JB, Griesinger C, Zweckstetter M (2009) EMBO J 28:3256
33. Sakellariou D, Brown SP, Lesage A, Hediger S, Bardet M, Meriles CA, Pines A, Emsley L (2003) J Am Chem Soc 125:4376
34. Mainz A, Jehle S, van Rossum BJ, Oschkinat H, Reif B (2009) J Am Chem Soc 131:15968
35. Bertini I, Luchinat C, Parigi G, Ravera E, Reif B, Turano P (2011) Proc Natl Acad Sci 108: 10396
36. Tycko R (2011) Annu Rev Phys Chem 62:279
37. Tycko R (2006) Q Rev Biophys 39:1
38. Goldbourt A, Gross BJ, Day LA, McDermott AE (2007) J Am Chem Soc 129:2338
39. Goldbourt A, Day LA, McDermott AE (2010) J Biol Chem 285:37051
40. Loquet A, Sgourakis NG, Gupta R, Giller K, Riedel D, Goosmann C, Griesinger C, Kolbe M, Baker D, Becker S, Lange A (2012) Nature 486:276
41. Hoyer W, Antony T, Cherny D, Heim G, Jovin TM, Subramaniam V (2002) J Mol Biol 322: 383
42. Petkova AT, Leapman RD, Guo Z, Yau W-M, Mattson MP, Tycko R (2005) Science 307:262
43. Paravastu AK, Leapman RD, Yau WM, Tycko R (2008) Proc Natl Acad Sci USA 105:18349
44. Qiang W, Yau WM, Tycko R (2011) J Am Chem Soc 133:4018
45. Paravastu AK, Qahwash I, Leapman RD, Meredith SC, Tycko R (2009) Proc Natl Acad Sci USA 106:7443
46. Kryndushkin DS, Wickner RB, Tycko R (2011) J Mol Biol 409:263
47. Cross TA, Sharma M, Yi M, Zhou HX (2011) Trends Biochem Sci 36:117
48. De Angelis AA, Opella SJ (2007) Nat Protoc 2:2332
49. Diller A, Loudet C, Aussenac F, Raffard G, Fournier S, Laguerre M, Grelard A, Opella SJ, Marassi FM, Dufourc EJ (2009) Biochimie 91:744
50. Cho HS, Dominick JL, Spence MM (2010) J Phys Chem B 114:9238
51. Grelard A, Loudet C, Diller A, Dufourc EJ (2010) Methods Mol Biol 654:341
52. Bayburt TH, Carlson JW, Sligar SG (1998) J Struct Biol 123:37

53. Li Y, Kijac AZ, Sligar SG, Rienstra CM (2006) Biophys J 91:3819
54. Kijac AZ, Li Y, Sligar SG, Rienstra CM (2007) Biochemistry 46:13696
55. Kijac A, Shih AY, Nieuwkoop AJ, Schulten K, Sligar SG, Rienstra CM (2010) Biochemistry 49:9190
56. Maricq MM, Waugh JS (1979) J Chem Phys 70:3300
57. Bajaj VS, van der Wel PC, Griffin RG (2009) J Am Chem Soc 131:118
58. Franks WT, Zhou DH, Wylie BJ, Money BG, Graesser DT, Frericks HL, Sahota G, Rienstra CM (2005) J Am Chem Soc 127:12291
59. Linden A, Franks W, Akbey Ü, Lange S, van Rossum B-J, Oschkinat H (2011) J Biomol NMR 51:283
60. Thurber KR, Tycko R (2008) J Magn Reson 195:179
61. Bertini I, Emsley L, Lelli M, Luchinat C, Mao J, Pintacuda G (2010) J Am Chem Soc 132: 5558
62. Laage S, Sachleben JR, Steuernagel S, Pierattelli R, Pintacuda G, Emsley L (2009) J Magn Reson 196:133
63. Lee M, Goldburg I (1965) Phys Rev 140:A1261
64. Dvinskikh SV, Zimmermann H, Maliniak A, Sandstrom D (2005) J Chem Phys 122:44512
65. Vinogradov E, Madhu PK, Vega S (1999) Chem Phys Lett 314:443
66. Sakellariou D, Lesage A, Hodgkinson P, Emsley L (2000) Chem Phys Lett 319:253
67. Haeberlen U, Waugh JS (1968) Phys Rev 175:453
68. Gerstein BC, Pembleton RG, Wilson RC, Ryan LM (1977) J Chem Phys 66:361
69. Vinogradov E, Madhu PK, Vega S (2002) Chem Phys Lett 354:193
70. Lesage A, Sakellariou D, Hediger S, Elena B, Charmont P, Steuernagel S, Emsley L (2003) J Magn Reson 163:105
71. Schaefer J, Stejskal EO (1976) J Am Chem Soc 98:1031
72. Bennett AE, Rienstra CM, Auger M, Lakshmi KV, Griffin RG (1995) J Chem Phys 103:6951
73. Detken A, Hardy EH, Ernst M, Meier BH (2002) Chem Phys Lett 356:298
74. Fung BM, Khitrin AK, Ermolaev K (2000) J Magn Reson 142:97
75. Ernst M, Samoson A, Meier BH (2001) Chem Phys Lett 348:293
76. Ladizhansky V (2009) Solid State Nucl Magn Reson 36:119
77. De Paëpe G (2012) Annu Rev Phys Chem 63:661
78. Gullion T, Schaefer J (1989) J Magn Reson 81:196
79. Hing AW, Vega S, Schaefer J (1992) J Magn Reson 96:205
80. Jaroniec CP, Tounge BA, Herzfeld J, Griffin RG (2001) J Am Chem Soc 123:3507
81. Baldus M, Petkova AT, Herzfeld J, Griffin RG (1998) Mol Phys 95:1197
82. Schütz A, Wasmer C, Habenstein B, Verel R, Greenwald J, Riek R, Böckmann A, Meier BH (2010) Chembiochem 11:1543
83. Andrew ER, Clough S, Farnell LF, Gledhill TD, Roberts I (1966) Phys Lett 21:505
84. Raleigh DP, Levitt MH, Griffin RG (1988) Chem Phys Lett 146:71
85. Levitt MH, Oas TG, Griffin RG (1988) Isr J Chem 28:271
86. Nielsen NC, Bildsoe H, Jakobsen HJ, Levitt MH (1994) J Chem Phys 101:1805
87. Levitt MH (2002) In: Harris RK (ed) Encyclopedia of nuclear magnetic resonance: supplementary volume. Wiley, Chichester, p 165
88. Hohwy M, Rienstra CM, Griffin RG (2002) J Chem Phys 117:4973
89. Bayro MJ, Huber M, Ramachandran R, Davenport TC, Meier BH, Ernst M, Griffin RG (2009) J Chem Phys 130:114506
90. Hohwy M, Rienstra CM, Jaroniec CP, Griffin RG (1999) J Chem Phys 110:7983
91. Bayro MJ, Maly T, Birkett NR, Dobson CM, Griffin RG (2009) Angew Chem Int Ed 48:5708
92. Caravatti P, Neuenschwander P, Ernst RR (1985) Macromolecules 18:119
93. Brus J, Petrickova H, Dybal J (2003) Solid State Nucl Magn Reson 23:183
94. Brown SP (2007) Prog Nucl Magn Reson Spectrosc 50:199
95. Zhang S, Meier BH, Ernst RR (1992) Phys Rev Lett 69:2149
96. Lange A, Luca S, Baldus M (2002) J Am Chem Soc 124:9704

97. Grommek A, Meier BH, Ernst M (2006) Chem Phys Lett 427:404
98. Castellani F, van Rossum B, Diehl A, Schubert M, Rehbein K, Oschkinat H (2002) Nature 420:98
99. Manolikas T, Herrmann T, Meier BH (2008) J Am Chem Soc 130:3959
100. Takegoshi K, Nakamura S, Terao T (2001) Chem Phys Lett 344:631
101. Takegoshi K, Nakamura S, Terao T (2003) J Chem Phys 118:2325
102. Morcombe CR, Gaponenko V, Byrd RA, Zilm KW (2004) J Am Chem Soc 126:7196
103. Weingarth M, Demco DE, Bodenhausen G, Tekely P (2009) Chem Phys Lett 469:342
104. Scholz I, Huber M, Manolikas T, Meier BH, Ernst M (2008) Chem Phys Lett 460:278
105. De Paepe G, Lewandowski JR, Loquet A, Bockmann A, Griffin RG (2008) J Chem Phys 129: 245101
106. Lewandowski JR, De Paepe G, Griffin RG (2007) J Am Chem Soc 129:728
107. Morris GA, Freeman R (1979) J Am Chem Soc 101:760
108. Andronesi OC, Becker S, Seidel K, Heise H, Young HS, Baldus M (2005) J Am Chem Soc 127:12965
109. Siemer AB, Arnold AA, Ritter C, Westfeld T, Ernst M, Riek R, Meier BH (2006) J Am Chem Soc 128:13224
110. Heise H, Hoyer W, Becker S, Andronesi OC, Riedel D, Baldus M (2005) Proc Natl Acad Sci USA 102:15871
111. Helmus JJ, Surewicz K, Surewicz WK, Jaroniec CP (2010) J Am Chem Soc 132:2393
112. Munowitz MG, Griffin RG, Bodenhausen G, Huang TH (1981) J Am Chem Soc 103:2529
113. Heise H, Celej MS, Becker S, Riede D, Pelah A, Kumar A, Jovin TM, Baldus M (2008) J Mol Biol 380:444
114. Helmus JJ, Surewicz K, Nadaud PS, Surewicz WK, Jaroniec CP (2008) Proc Natl Acad Sci USA 105:6284
115. Böckmann A, Meier B (2010) Prion 4:72
116. Glenner GG, Wong CW (1984) Biochem Biophys Res Commun 120:885
117. Kang J, Lemaire H-G, Unterbeck A, Salbaum JM, Masters CL, Grzeschik K-H, Multhaup G, Beyreuther K, Muller-Hill B (1987) Nature 325:733
118. Tycko R, Sciarretta KL, Orgel JP, Meredith SC (2009) Biochemistry 48:6072
119. Lopez del Amo JM, Schmidt M, Fink U, Dasari M, Fändrich M, Reif B (2012) Angew Chem Int Ed 51:6136
120. Bertini I, Gonnelli L, Luchinat C, Mao J, Nesi A (2011) J Am Chem Soc 133:16013
121. Chimon S, Shaibat MA, Jones CR, Calero DC, Aizezi B, Ishii Y (2007) Nat Struct Mol Biol 14:1157
122. Ahmed M, Davis J, Aucoin D, Sato T, Ahuja S, Aimoto S, Elliott JI, Van Nostrand WE, Smith SO (2010) Nat Struct Mol Biol 17:561
123. Scheidt HA, Morgado I, Rothemund S, Huster D, Fandrich M (2011) Angew Chem Int Ed Engl 50:2837
124. Scheidt HA, Morgado I, Huster D (2012) J Biol Chem 287:22822
125. Spillantini MG, Schmidt ML, Lee VMY, Trojanowski JQ, Jakes R, Goedert M (1997) Nature 388:839
126. Kruger R, Kuhn W, Muller T, Woitalla D, Graeber M, Kosel S, Przuntek H, Epplen JT, Schols L, Riess O (1998) Nat Genet 18:106
127. Polymeropoulos MH, Lavedan C, Leroy E, Ide SE, Dehejia A, Dutra A, Pike B, Root H, Rubenstein J, Boyer R, Stenroos ES, Chandrasekharappa S, Athanassiadou A, Papapetropoulos T, Johnson WG, Lazzarini AM, Duvoisin RC, DiIorio G, Golbe LI, Nussbaum RL (1997) Science 276:2045
128. Zarranz JJ, Alegre J, Gomez-Esteban JC, Lezcano E, Ros R, Ampuero I, Vidal L, Hoenicka J, Rodriguez O, Atares B, Llorens V, Tortosa EG, del Ser T, Munoz DG, de Yebenes JG (2004) Ann Neurol 55:164
129. Kloepper KD, Woods WS, Winter KA, George JM, Rienstra CM (2006) Protein Expr Purif 48:112

130. Kloepper K, Zhou D, Li Y, Winter K, George J, Rienstra C (2007) J Biomol NMR 39:197
131. Kloepper KD, Hartman KL, Ladror DT, Rienstra CM (2007) J Phys Chem B 111:13353
132. Vilar M, Chou HT, Luhrs T, Maji SK, Riek-Loher D, Verel R, Manning G, Stahlberg H, Riek R (2008) Proc Natl Acad Sci USA 105:8637
133. Loquet A, Giller K, Becker S, Lange A (2010) J Am Chem Soc 132:15164
134. Gath J, Habenstein B, Bousset L, Melki R, Meier B, Böckmann A (2012) Biomol NMR Assign 6:51
135. Chen M, Margittai M, Chen J, Langen R (2007) J Biol Chem 282:24970
136. Lemkau LR, Comellas G, Kloepper KD, Woods WS, George JM, Rienstra CM (2012) J Biol Chem 287:11526
137. Comellas G, Lemkau LR, Nieuwkoop AJ, Kloepper KD, Ladror DT, Ebisu R, Woods WS, Lipton AS, George JM, Rienstra CM (2011) J Mol Biol 411:881
138. Saibil HR, Seybert A, Habermann A, Winkler J, Eltsov M, Perkovic M, Castaño-Diez D, Scheffer MP, Haselmann U, Chlanda P, Lindquist S, Tyedmers J, Frangakis AS (2012) Proc Natl Acad Sci USA 109:14906
139. Wickner RB (1994) Science 264:566
140. Suzuki G, Shimazu N, Tanaka M (2012) Science 336:355
141. Shewmaker F, Wickner RB, Tycko R (2006) Proc Natl Acad Sci USA 103:19754
142. Baxa U, Wickner RB, Steven AC, Anderson DE, Marekov LN, Yau WM, Tycko R (2007) Biochemistry 46:13149
143. Wickner RB, Dyda F, Tycko R (2008) Proc Natl Acad Sci USA 105:2403
144. Shewmaker F, Kryndushkin D, Chen B, Tycko R, Wickner RB (2009) Biochemistry 48:5074
145. Chan JCC, Oyler NA, Yau WM, Tycko R (2005) Biochemistry 44:10669
146. Goncalves JA, Ahuja S, Erfani S, Eilers M, Smith SO (2010) Prog Nucl Magn Reson Spectrosc 57:159
147. Stouffer AL, Acharya R, Salom D, Levine AS, Di Costanzo L, Soto CS, Tereshko V, Nanda V, Stayrook S, DeGrado WF (2008) Nature 451:596
148. van der Wel PCA, Lewandowski JR, Griffin RG (2007) J Am Chem Soc 129:5117
149. Lewandowski JR, van der Wel PCA, Rigney M, Grigorieff N, Griffin RG (2011) J Am Chem Soc 133:14686
150. Loquet A, Bousset L, Gardiennet C, Sourigues Y, Wasmer C, Habenstein B, Schütz A, Meier BH, Melki R, Böckmann A (2009) J Mol Biol 394:108
151. Ritter C, Maddelein ML, Siemer AB, Luhrs T, Ernst M, Meier BH, Saupe SJ, Riek R (2005) Nature 435:844
152. Wasmer C, Lange A, Van Melckebeke H, Siemer AB, Riek R, Meier BH (2008) Science 319:1523
153. Van Melckebeke H, Wasmer C, Lange A, Ab E, Loquet A, Böckmann A, Meier BH (2010) J Am Chem Soc 132:13765
154. Wasmer C, Schütz A, Loquet A, Buhtz C, Greenwald J, Riek R, Böckmann A, Meier BH (2009) J Mol Biol 394:119
155. Schütz AK, Soragni A, Hornemann S, Aguzzi A, Ernst M, Böckmann A, Meier BH (2011) Angew Chem Int Ed 50:5956
156. Hong M, Zhang Y, Hu FH (2012) Annu Rev Phys Chem 63:1
157. Smith SO (2010) Annu Rev Biophys 39:309
158. Schnell JR, Chou JJ (2008) Nature 451:591
159. Cady SD, Hong M (2008) Proc Natl Acad Sci USA 105:1483
160. Cady SD, Mishanina TV, Hong MJ (2009) Mol Biol 385:1127
161. Cady SD, Schmidt-Rohr K, Wang J, Soto CS, DeGrado WF, Hong M (2010) Nature 463:689
162. Cady S, Wang T, Hong M (2011) J Am Chem Soc 133:11572
163. Miao Y, Qin H, Fu R, Sharma M, Can TV, Hung I, Luca S, Gor'kov PL, Brey WW, Cross TA (2012) Angew Chem Int Ed 51:8383
164. Hu FH, Luo WB, Hong M (2010) Science 330:505
165. Hong M, Fritzsching KJ, Williams JK (2012) J Am Chem Soc 134:14753

166. Luo W, Hong M (2010) J Am Chem Soc 132:2378
167. Andreas LB, Eddy MT, Chou JJ, Griffin RG (2012) J Am Chem Soc 134:7215
168. Oesterhelt D, Stoeckenius W (1971) Nature 233:149
169. Béjà O, Aravind L, Koonin EV, Suzuki MT, Hadd A, Nguyen LP, Jovanovich SB, Gates CM, Feldman RA, Spudich JL, Spudich EN, DeLong EF (2000) Science 289:1902
170. Gordeliy VI, Labahn J, Moukhametzianov R, Efremov R, Granzin J, Schlesinger R, Buldt G, Savopol T, Scheidig AJ, Klare JP, Engelhard M (2002) Nature 419:484
171. Gautier A, Mott HR, Bostock MJ, Kirkpatrick JP, Nietlispach D (2010) Nat Struct Mol Biol 17:768
172. Reckel S, Gottstein D, Stehle J, Löhr F, Verhoefen M-K, Takeda M, Silvers R, Kainosho M, Glaubitz C, Wachtveitl J, Bernhard F, Schwalbe H, Güntert P, Dötsch V (2011) Angew Chem Int Ed 50:11942
173. Klare JP, Bordignon E, Doebber M, Fitter J, Kriegsmann J, Chizhov I, Steinhoff HJ, Engelhard M (2006) J Mol Biol 356:1207
174. Moukhametzianov R, Klare JP, Efremov R, Baeken C, Goppner A, Labahn J, Engelhard M, Buldt G, Gordeliy VI (2006) Nature 440:115
175. Etzkorn M, Seidel K, Li L, Martell S, Geyer M, Engelhard M, Baldus M (2010) Structure 18:293
176. Bajaj VS, Mak-Jurkauskas ML, Belenky M, Herzfeld J, Griffin RG (2009) Proc Natl Acad Sci USA 106:9244
177. Mak-Jurkauskas ML, Bajaj VS, Hornstein MK, Belenky M, Griffin RG, Herzfeld J (2008) Proc Natl Acad Sci USA 105:883
178. Harbison GS, Smith SO, Pardoen JA, Winkel C, Lugtenburg J, Herzfeld J, Mathies R, Griffin RG (1984) Proc Natl Acad Sci USA 81:1706
179. Jaroniec CP, Lansing JC, Tounge BA, Belenky M, Herzfeld J, Griffin RG (2001) J Am Chem Soc 123:12929
180. Higman VA, Varga K, Aslimovska L, Judge PJ, Sperling LJ, Rienstra CM, Watts A (2011) Angew Chem Int Ed Engl 50(36):8432
181. Shi L, Ahmed MAM, Zhang W, Whited G, Brown LS, Ladizhansky V (2009) J Mol Biol 386:1078
182. Shi L, Kawamura I, Jung K-H, Brown LS, Ladizhansky V (2011) Angew Chem Int Ed 50:1302
183. Wang S, Shi L, Okitsu T, Wada A, Brown LS, Ladizhansky V (2012) Biomol NMR Assign
184. Pfleger N, Worner AC, Yang J, Shastri S, Hellmich UA, Aslimovska L, Maier MS, Glaubitz C (2009) Biochim Biophys Acta 1787:697
185. Yang J, Aslimovska L, Glaubitz C (2011) J Am Chem Soc 133:4874
186. Pfleger N, Lorch M, Woerner AC, Shastri S, Glaubitz C (2008) J Biomol NMR 40:15
187. Ward ME, Shi L, Lake E, Krishnamurthy S, Hutchins H, Brown LS, Ladizhansky VJ (2011) J Am Chem Soc 133:17434
188. Hempelmann F, Holper S, Verhoefen MK, Woerner AC, Kohler T, Fiedler SA, Pfleger N, Wachtveitl J, Glaubitz C (2011) J Am Chem Soc 133:4645
189. Wang S, Munro RA, Kim SY, Jung KH, Brown LS, Ladizhansky V (2012) J Am Chem Soc 134:16995
190. Hopkins AL, Groom CR (2002) Nat Rev Drug Discov 1:727
191. Jacoby E, Bouhelal R, Gerspacher M, Seuwen K (2006) ChemMedChem 1:761
192. Klabunde T, Hessler G (2002) Chembiochem 3:928
193. Rosenbaum DM, Rasmussen SG, Kobilka BK (2009) Nature 459:356
194. Zhao Q, Wu BL (2012) Acta Pharmacol Sin 33:324
195. Katritch V, Cherezov V, Stevens RC (2013) Annu Rev Pharmacol Toxicol 53:531
196. Kobilka B, Schertler GFX (2008) Trends Pharmacol Sci 29:79
197. Tapaneeyakorn S, Goddard AD, Oates J, Willis CL, Watts A (2011) Biochim Biophys Acta 1808:1462

198. Ahuja S, Hornak V, Yan ECY, Syrett N, Goncalves JA, Hirshfeld A, Ziliox M, Sakmar TP, Sheves M, Reeves PJ, Smith SO, Eilers M (2009) Nat Struct Mol Biol 16:168
199. Park SH, Prytulla S, De Angelis AA, Brown JM, Kiefer H, Opella SJ (2006) J Am Chem Soc 128:7402
200. Goncalves JA, South K, Ahuja S, Zaitseva E, Opefi CA, Eilers M, Vogel R, Reeves PJ, Smith SO (2010) Proc Natl Acad Sci USA 107:19861
201. Struts AV, Salgado GF, Martinez-Mayorga K, Brown MF (2011) Nat Struct Mol Biol 18:392
202. Mertz B, Struts AV, Feller SE, Brown MF (2012) Biochim Biophys Acta 1818:241
203. Overhauser AW (1953) Phys Rev 92:411
204. Carver TR, Slichter CP (1953) Phys Rev 92:212
205. Carver TR, Slichter CP (1956) Phys Rev 102:975
206. Hu KN, Debelouchina GT, Smith AA, Griffin RG (2011) J Chem Phys 134:125105
207. Smith AA, Corzilius B, Barnes AB, Maly T, Griffin RG (2012) J Chem Phys 136:015101
208. Wind RA, Duijvestijn MJ, Vanderlugt C, Manenschijn A, Vriend J (1985) Prog Nucl Magn Reson Spectrosc 17:33
209. Wollan DS (1976) Phys Rev B 13:3671
210. Maly T, Debelouchina GT, Bajaj VS, Hu KN, Joo CG, Mak-Jurkauskas ML, Sirigiri JR, van der Wel PCA, Herzfeld J, Temkin RJ, Griffin RG (2008) J Chem Phys 128:052211
211. Felch KL, Danly BG, Jory HR, Kreischer KE, Lawson W, Levush B, Temkin RJ (1999) Proc IEEE 87:752
212. Becerra LR, Gerfen GJ, Temkin RJ, Singel DJ, Griffin RG (1993) Phys Rev Lett 71:3561
213. Becerra LR, Gerfen GJ, Bellew BF, Bryant JA, Hall DA, Inati SJ, Weber RT, Un S, Prisner TF, McDermott AE, Fishbein KW, Kreischer KE, Temkin RJ, Singel DJ, Griffin RG (1995) J Magn Reson Ser A 117:28
214. Haze O, Corzilius B, Smith AA, Griffin RG, Swager TM (2012) J Am Chem Soc 134:14287
215. Hu KN, Yu HH, Swager TM, Griffin RG (2004) J Am Chem Soc 126:10844
216. Song CS, Hu KN, Joo CG, Swager TM, Griffin RG (2006) J Am Chem Soc 128:11385
217. Matsuki Y, Maly T, Ouari O, Karoui H, Le Moigne F, Rizzato E, Lyubenova S, Herzfeld J, Prisner T, Tordo P, Griffin RG (2009) Angew Chem Int Ed Engl 48:4996
218. Matsuki Y, Takahashi H, Ueda K, Idehara T, Ogawa I, Toda M, Akutsu H, Fujiwara T (2010) Phys Chem Chem Phys 12:5799
219. Hall DA, Maus DC, Gerfen GJ, Inati SJ, Becerra LR, Dahlquist FW, Griffin RG (1997) Science 276:930
220. Rosay M, Zeri AC, Astrof NS, Opella SJ, Herzfeld J, Griffin RG (2001) J Am Chem Soc 123:1010
221. vanderWel PCA, Hu KN, Lewandowski J, Griffin RG (2006) J Am Chem Soc 128:10840
222. Bajaj VS, Mak-Jurkauskas ML, Belenky M, Herzfeld J, Griffin RG (2010) J Magn Reson 202:9
223. Akbey U, Linden AH, Oschkinat H (2012) Appl Magn Reson 43:81
224. Barnes AB, Corzilius B, Mak-Jurkauskas ML, Andreas LB, Bajaj VS, Matsuki Y, Belenky ML, Lugtenburg J, Sirigiri JR, Temkin RJ, Herzfeld J, Griffin RG (2010) Phys Chem Chem Phys 12: 5861
225. Lange S, Linden AH, Akbey Ü, Trent Franks W, Loening NM, Rossum B-J, Oschkinat H (2012) J Magn Reson 216:209
226. Debelouchina GT, Bayro MJ, van der Wel PCA, Caporini MA, Barnes AB, Rosay M, Maas WE, Griffin RG (2010) Phys Chem Chem Phys 12:5911
227. Bayro MJ, Debelouchina GT, Eddy MT, Birkett NR, MacPhee CE, Rosay M, Maas WE, Dobson CM, Griffin RG (2011) J Am Chem Soc 133:13967
228. Sergeyev IV, Day LA, Goldbourt A, McDermott AE (2011) J Am Chem Soc 133:20208
229. Reggie L, Lopez JJ, Collinson I, Glaubitz C, Lorch M (2011) J Am Chem Soc 133:19084
230. Linden AH, Lange S, Franks WT, Akbey Ü, Specker E, van Rossum B-J, Oschkinat H (2011) J Am Chem Soc 133:19266

231. Krabben L, van Rossurn BJ, Jehle S, Bocharov E, Lyukmanova EN, Schulga AA, Arseniev A, Hucho F, Oschkinat H (2009) J Mol Biol 390:662
232. Jacso T, Franks WT, Rose H, Fink U, Broecker J, Keller S, Oschkinat H, Reif B (2012) Angew Chem Int Ed 51:432
233. Renault M, Pawsey S, Bos MP, Koers EJ, Nand D, Tommassen-van Boxtel R, Rosay M, Tommassen J, Maas WE, Baldus M (2012) Angew Chem Int Ed 51:2998
234. Maly T, Cui DT, Griffin RG, Miller AF (2012) J Phys Chem B 116:7055
235. Corzilius B, Smith AA, Barnes AB, Luchinat C, Bertini I, Griffin RG (2011) J Am Chem Soc 133:5648

Top Curr Chem (2013) 335: 157–200
DOI: 10.1007/128_2011_312

Published online: 6 March 2012

Paramagnetic Solid-State Magic-Angle Spinning NMR Spectroscopy

Guido Pintacuda and Gwendal Kervern

Abstract A number of technical improvements have recently opened up solid-state NMR to the analysis of new classes of substrates with wide ranging implications for molecular and biological sciences, with an immediate impact on a large community of researchers. A wealth of information can be extracted from the analysis of solid-state NMR signals of paramagnetic compounds, as the changes induced by the paramagnetic center depend in a well-defined way on the structure of the molecule. Solid-state NMR is in a position to allow direct, straightforward experimental access to the fine details of the molecular electronic configuration, which is in turn a sensible reporter of the molecular geometry in small catalysts as well as in larger biomolecules.

Keywords Hyperfine coupling · Magic angle spinning · Paramagnetism · Solid-state NMR

Contents

G. Pintacuda (✉)
Centre de RMN à Très Hauts Champs, Institut de Sciences Analytiques, Université de Lyon, UMR 5280 CNRS/Ecole Normale Supérieure de Lyon/Université Claude Bernard Lyon 1, 69100 Villeurbanne, France
e-mail: guido.pintacuda@ens-lyon.fr

G. Kervern
Centre de RMN à Très Hauts Champs, Institut de Sciences Analytiques, Université de Lyon, UMR 5280 CNRS/Ecole Normale Supérieure de Lyon/Université Claude Bernard Lyon 1, 69100 Villeurbanne, France

Present address: Université Henri Poincaré (UHP), Nancy-Université, Faculté des Sciences et Techniques, 54506 Vandœuvre-lès-Nancy

1 Introduction

High-resolution solid-state NMR using magic angle spinning (MAS) is a powerful technique for characterization and structural analysis of diamagnetic systems in solids [1, 2], including noncrystalline organic materials [3], surface catalysts [4], and biomolecules [5].

Many samples of interest in chemistry and in biology, however, contain paramagnetic centers. Paramagnetic centers originate from unpaired electrons that are intrinsic features of organic radicals and of many transition metal ions. A paramagnetic center interacts with the surrounding nuclear spins and changes the appearance of their NMR spectrum by altering their chemical shifts and increasing their relaxation rates [6]. The effects arising from paramagnetism depend in a well-defined manner on the electronic configuration of the metal and on the structure of the molecule, providing in principle a variety of information on the electronic states of the paramagnetic center, as well as a number of structural restraints in the surrounding environment [7, 8]. In contrast to diamagnetic data, paramagnetic phenomena provide long-range information due to the large value of the electron magnetic moment.

Solution-state NMR studies of paramagnetic systems are well-developed, and the use of paramagnetic restraints for protein structure determination has been amply demonstrated [9]. In solids, the large spectral dispersion due to the large orientation-dependent hyperfine shifts, as well as the paramagnetically-enhanced nuclear relaxation, have traditionally imposed severe technical difficulties in high-resolution solid-state NMR studies, and paramagnetic materials have long only been the object of low resolution NMR in solid-state physics [10].

The first reports on MAS NMR of paramagnetic molecular solids date back to 1986, when Bryant and coworkers [11] obtained ^{13}C CP MAS spectra of polycrystalline Pr^{III} and Eu^{III} acetates, and when Haw and coworkers [12, 13] showed that the ^{13}C lines of Sm^{III} acetate move in a CP MAS spectrum according to a Curie law behavior, proposing ^{13}C CP MAS of this compound as the basis for a solid-state NMR chemical-shift thermometer.

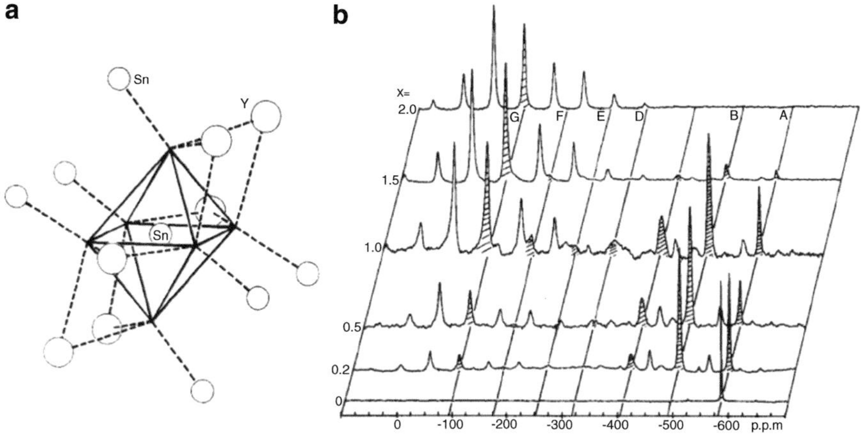

Fig. 1 (**a**) The local environment of the tin in $Y_2Sn_2O_7$ showing the next-nearest tin and yttrium ion neighbors. (**b**) ^{119}Sn MAS NMR spectra from samples of variable stoichiometries $Y_{2-x}Sm_xSn_2O_7$, $x = 0.0$ ($Y_2Sn_2O_7$), 0.2, 0.5, 1.0, 1.5, and 2.0 ($Sm_2Sn_2O_7$) (field of 4.7 T, 3.5 kHz MAS). Isotropic resonances of the different local environments (A–G) are *shaded* and each is traced as a function of the composition, while the remaining peaks are *sidebands*. Reprinted by permission from Macmillan Publishers Ltd: Nature Publishing Group [14]. © 1987

The first systematic analysis of paramagnetic effects in solid-state NMR was provided at the end of the 1980s by a series of works by Clare Grey, Chris Dobson and coworkers on ^{119}Sn and ^{89}Y NMR of several pyrochlore solid solutions $Y_{2-x}Ln_xSn_2O_7$ (Ln=La, Pr, Nd, Sm, Eu, Tm, Yb, Lu) [14–16]. These works show that it is possible to obtain and to interpret high-resolution NMR spectra from solids containing paramagnetic lanthanides. Besides the obvious potential in systems where the distributions of rare-earth ions is of importance for understanding the properties of the materials involved (in laser materials, phosphors, and host materials for fission product elements), the approach outlined in these papers has become a cornerstone for the use of paramagnetic shifts in MAS NMR to probe the structures of continuous solids, as well as of powdered molecular complexes (Fig. 1).

In the solutions at different Ln^{III} concentrations, in addition to the resonance of the diamagnetic end-member compound $Y_2Sn_2O_7$, a series of extra ^{119}Sn and ^{89}Y peaks were observed, associated with the substitution of paramagnetic for diamagnetic lanthanide ions in the primary coordination sphere around a tin or a yttrium atom. The chemical shifts of the ^{119}Sn and ^{89}Y nuclei were found to be extremely sensitive to the adjacent paramagnetic lanthanide ions, each successive substitution of Ln^{III} for Y^{III} into the local coordination sphere around the tin producing an additive shift, as well as a decrease in the spin-lattice relaxation time (T_1). Notably, the paramagnetic end-point spectra ($Ln_2Sn_2O_7$) exhibited a very large variation in ^{119}Sn chemical shifts (from approximately +5,400 to −4,200 ppm), and had large overall linewidths associated with the substantial anisotropy of the shift, but the individual peaks within the spinning sideband manifolds remained sharp. The intensities of these peaks allowed the determination of the composition of the solid solution, which had proved impossible from powder X-ray diffraction data.

These results exemplify *in a nutshell* advantages and issues of paramagnetic solid-state MAS NMR: substitution of diamagnetic cations with paramagnetic cations leads to (a) massive chemical shifts, which provide at the same time increased spectral resolution and structural information on the system, (b) large shift anisotropies, which encode long-range nuclear-electron distances, and (c) enhanced longitudinal relaxation, with the possibility of observing fast relaxing nuclei despite their low abundance, by varying the recycle times between pulses and saturating selectively the resonances of nuclei far from the paramagnetic ions.

2 A Bit of Theory

The Hamiltonian to consider for the theoretical analysis of the NMR properties of nuclei in an open-shell system contains a field- and nuclear spin-free electronic part $\mathcal{H}_0^S$, the nuclear and electron Zeeman parts $\mathcal{H}_Z^I$ and $\mathcal{H}_Z^S$, and an interaction term $\mathcal{H}_{int}$:

$$\mathcal{H} = \mathcal{H}_0^S + \mathcal{H}_Z^S + \mathcal{H}_Z^I + \mathcal{H}_{int}. \tag{1}$$

The field- and nuclear spin-free electronic part $\mathcal{H}_0^S$ contains the kinetic energy and electrostatic interactions involving electrons and (fixed) nuclei.

The Zeeman terms $\mathcal{H}_Z^I$ and $\mathcal{H}_Z^S$ are respectively

$$\mathcal{H}_Z^I = \hbar\gamma_I \mathrm{I.B_0} \text{ and } \mathcal{H}_Z^S = \mu_B[\mathcal{L} + \mathrm{g_e S}].\mathrm{B_0}, \tag{2}$$

where I and S are the nuclear and electron spin operators, L is the electron orbital operator, γ_I the nuclear gyromagnetic moment, μ_B the Bohr magneton, and $g_e = 2.003$ the free-electron Landé g factor defined after the electronic gyromagnetic moment γ_S: $\hbar\gamma_S = \mu_B\, g_e$.

Finally, $\mathcal{H}_{int}$ describes the electron–nucleus hyperfine coupling. A number of important contributions affect this term, such as those arising, for example, from g-tensor anisotropy, zero-field splitting, and spin-orbit coupling. However, profound gaps in the basic underlying physical formalism, and the lack of quantitative quantum-mechanical treatments of the parameters of paramagnetic NMR, have for a long time rendered the expression of $\mathcal{H}_{int}$ difficult to treat. While only recently a framework has been developed for the description and calculation of paramagnetic effects from first-principles [17], notably thanks to the efforts carried out by Kaupp, Malkin, Vaara and their coworkers (see, e.g., [18–20]), the interpretation of the NMR spectra obtained in the case of paramagnetic metal centers has required an intermediate descriptor, the magnetic susceptibility tensor.

Magnetic susceptibility. The so-called magnetic susceptibility ("χ") is the average electronic moment $\langle \mu_s \rangle$ of the electron spin in the magnetic field B_0:

$$\langle \mu_s \rangle = \frac{\chi}{\mu_0} B_0, \tag{3}$$

where μ_0 is the vacuum magnetic permeability.

For a given electron spin quantum number S, the Boltzmann population of each electron Zeeman state, defined by the magnetic quantum number M_S, with respect to the lowest Zeeman state is

$$\frac{N_{M_S}}{N_0} = \frac{\exp\left\{\frac{-g_e \mu_B M_S B_0}{k_B T}\right\}}{\sum_{M_S} \exp\left\{\frac{-g_e \mu_B M_S B_0}{k_B T}\right\}}, \tag{4}$$

where μ_B is the Bohr magneton, g_e the electron g-factor, k_B the Boltzmann constant, and T the absolute temperature. As the dynamics of electron relaxation and precession is several orders of magnitude faster than the NMR timescale, the electron spin S detected in the NMR experiment reduces to the so-called "Curie spin" $\langle S_z \rangle$ averaged over the different Zeeman states:

$$\langle S_z \rangle = \frac{\sum_{M_S} M_S \exp\left\{\frac{-g_e \mu_B M_S B_0}{k_B T}\right\}}{\sum_{M_S} \exp\left\{\frac{-g_e \mu_B M_S B_0}{k_B T}\right\}}, \tag{5}$$

$$\langle S_\pm \rangle = 0. \tag{6}$$

In the high-temperature approximation, (5) simplifies to

$$\langle S_z \rangle = -\frac{g_e \mu_B B_0}{3 k_B T} S(S+1), \tag{7}$$

which yields for the average moment $\langle \mu_S \rangle$ of this "Curie spin":

$$\langle \mu_S \rangle = \frac{g_e^2 \mu_B^2 S(S+1)}{3 k_B T} B_0 \tag{8}$$

From (7) we obtain for the molecular magnetic susceptibility the following expression (Curie law):

$$\chi_{iso} = \mu_0 \frac{g_e^2 \mu_B^2 S(S+1)}{3 k_B T}, \tag{9}$$

and the Curie moment (8) can be expressed as

$$\langle S\rangle = \langle S_z\rangle = -\frac{\chi_{\mathrm{iso}}}{\mu_0\mu_{\mathrm{B}}g_{\mathrm{e}}}B_0. \tag{10}$$

Thus, the total electron spin is sensed by the surrounding nuclei as an additional magnetic moment which is parallel and proportional to the external magnetic field. The effect of coupling between electrons and nuclei results, therefore, in a shift rather than a splitting.

For orbitally degenerate systems, in the presence of zero-field splitting or strong spin-orbit coupling ($Fe^{II/III}$, Co^{II}, Ln^{III}, . . .), the expression of the magnetic susceptibility becomes a symmetric rank-2 tensor χ, which can be typically defined empirically. In its principal axis system (PAS), i.e., the frame in which the tensor is diagonal (with $|\chi_{zz}| \geq |\chi_{yy}| \geq |\chi_{xx}|$), χ is defined by its isotropic value χ_{iso}, its anisotropy $\Delta\chi$, and its asymmetry η:

$$\begin{aligned}
\Delta\chi &= \chi_{zz} - \chi_{\mathrm{iso}} \\
\chi_{\mathrm{iso}} &= \frac{1}{3}\sum_i \chi_{ii} = \frac{1}{3}Tr(\chi) \\
\eta &= \frac{\chi_{xx} - \chi_{yy}}{\Delta\chi},
\end{aligned} \tag{11}$$

or, alternatively, by its axial and rhombic components:

$$\Delta\chi_{\mathrm{ax}} = \chi_{zz} - \frac{\chi_{xx} + \chi_{yy}}{2}, \quad \Delta\chi_{\mathrm{rh}} = \chi_{xx} - \chi_{yy}, \tag{12}$$

or again, in spherical components:

$$\begin{aligned}
\chi_{0,0} &= -\sqrt{3}\chi_{\mathrm{iso}} \\
\chi_{2,0} &= \sqrt{\frac{2}{3}}\Delta\chi_{\mathrm{ax}} \\
\chi_{2,\pm 2} &= \frac{3}{4}\frac{\Delta\chi_{\mathrm{rh}}}{\Delta\chi_{\mathrm{ax}}}.
\end{aligned} \tag{13}$$

Within the susceptibility formalism, the coupling between the nucleus and the average electronic moment is thus interpreted as the sum of two distinct interactions:

$$\mathcal{H}_{\mathrm{int}} = \mathcal{H}^{\mathrm{C}} + \mathcal{H}^{\mathrm{D}}, \tag{14}$$

that is (a) a through bond or Fermi contact effect $\mathcal{H}^{\mathrm{C}}$, caused by positive and negative spin densities on remote spins due to partial transfer of the unpaired electrons by direct delocalization and spin polarization, and (b) a through-space dipolar effect $\mathcal{H}^{D}$, resulting from the dipolar coupling between the average dipole moment of the electronic distribution and the surrounding nuclear spins.

Fermi-contact coupling. The through-bond coupling between the average electron spin and a nuclear spin I has a rank-0, isotropic component A (hyperfine Fermi constant) which is proportional to the spin density in the s orbital of the atom carrying the I spin:

$$\mathcal{H}^{C} = \mathcal{A}\chi \cdot \mathrm{I}, \tag{15}$$

where

$$\mathcal{A} = \frac{\mu_0}{3\mathrm{S}}\hbar g_e \mu_B \gamma_I \rho_s, \tag{16}$$

where γ_I and γ_S are the nuclear and the electronic gyromagnetic ratios, g_e is the electron g value, μ_B the Bohr magneton, and ρ_s can be obtained summing positive $\left|\psi_i^+(0)\right|^2$ and negative $\left|\psi_i^-(0)\right|^2$ spin densities at the nucleus for all the ith molecular orbitals:

$$\rho_s = \sum_i \left[\left|\psi_i^-(0)\right|^2 - \left|\psi_i^+(0)\right|^2\right]. \tag{17}$$

The isotropic Fermi contact shift δ^{C} in parts per million resulting from this interaction is

$$\delta^{C} = \frac{\mathcal{A}}{\hbar}\frac{\chi}{\mu_0 \gamma_I g_S \mu_B} = \frac{\mathcal{A}}{\hbar}\frac{g_e \mu_B S(S+1)}{3\gamma_I k_B T}. \tag{18}$$

Dipolar coupling. Within a point-dipole approximation, the "through-space" dipolar interaction between a nuclear spin and electron spin density outside the nucleus, $\mathcal{H}^{D}$, is expressed in tensor form as

$$\mathcal{H}^{D} = \sum_{m=-1,0,1} \mathcal{D}_{2,0}\, d_{0m}^2(\beta_{PL}) T_{2,m}, \tag{19}$$

where $\mathcal{D}_{2,0}$ is the only non-zero component of the traceless, axially symmetric rank-2 dipolar tensor $\mathcal{D}$:

$$\mathcal{D}_{2,0} = \sqrt{6}\frac{\hbar \gamma_I \mu_B g_e}{4\pi r_{IS}^3}, \tag{20}$$

where γ_I is the nuclear gyromagnetic ratio, g_e the electron g value, μ_B the Bohr magneton, r_{IS} the nucleus–electron distance, $d_{m,m'}^l(\beta_{PL})$ is the reduced Wigner matrix element, and β_{PL} is the angle between the principal axis of the $\mathcal{D}$ tensor and the magnetic field. Finally, the spin-operator terms $T_{2,m}$ result from a dyadic product of the χ tensor components and the nuclear spin operator I, respectively.

In a high magnetic field, the hyperfine Hamiltonian can be truncated after the dominant Zeeman interaction with the magnetic field (secular approximation). Only the terms in I_z are thus retained, and the spin-operators reduce to

$$\begin{aligned} T_{2,0} &= \frac{2}{3}\chi_{2,0}\, I_z B_0 - \frac{\sqrt{2}}{3}\chi_{0,0} I_z B_0 \\ T_{2,\pm1} &= \mp\frac{1}{2}\langle S_\pm\rangle I_z = 0 \\ T_{2,\pm2} &= 0. \end{aligned} \tag{21}$$

As many NMR applications require rotations or other unitary transformations, it is convenient to reformulate the dipolar interaction as a sum of terms $W_{N,0}$ of ranks 0, 1, 2, …N, as each of these terms behaves differently under rotation. Equation (19) can then be recast as (see Appendix for the full derivation):

$$\begin{aligned} \tilde{\mathcal{H}}^{\mathrm{D}} &= \{a_{0,0}W_{0,0} + a_{2,0}W_{2,0}\} I_z B_0 \\ &= \left\{\delta^{\mathrm{PC}} + \sqrt{\frac{2}{3}}\Delta\sigma^{\mathrm{D}} d^2_{0,0}(\beta_{\mathrm{PL}})\right\} I_z B_0 \\ &= \left\{\delta^{\mathrm{PC}} + \sqrt{\frac{1}{6}}\Delta\sigma^{\mathrm{D}}(3\cos^2\beta_{\mathrm{PL}} - 1)\right\} I_z B_0. \end{aligned} \tag{22}$$

The rank-zero interaction, δ^{PC} [the pseudocontact shift (PCS)] is the isotropic shift from the dipolar coupling of the nucleus with the electron, and the rank-two components represent a shift anisotropy analogous to the diamagnetic CSA (the "dipolar shift anisotropy" or DSA). It is interesting to note that the final form of the hyperfine interaction is in all respects analogous to that of a diamagnetic chemical shift. Notably, $W_{4,0}$ terms vanish, in contrast to, for example, the second-order quadrupolar effect where the product of two second-rank components contributes to the $W_{4,0}$ [21].

In the solid state, all possible crystallite orientations are present simultaneously in powder samples. Each crystallite is subject to a different perturbation throughout an NMR experiment due to the orientation dependence in the anisotropic part of the Hamiltonian, and so the final signal is the result of a powder sum over all relevant orientations. This form for the lineshape leads to the so-called *powder pattern*. In a point-dipole approximation, this shift pattern is characterized by an axially symmetric anisotropy $\Delta\sigma^{\mathrm{D}}$ [22, 23], which corresponds to the rank-2 terms contained in (22). This shift anisotropy can be expressed as

$$\Delta\sigma^{\mathrm{D}} = \sqrt{\frac{2}{3}}\, W_{2,0} \simeq \frac{\chi_{\mathrm{iso}}}{\hbar\gamma_{\mathrm{I}} r^3} = \frac{\mu_{\mathrm{o}}}{\hbar\gamma_{\mathrm{I}} r^3}\,\frac{g_{\mathrm{e}}^2\mu_{\mathrm{B}}^2 S(S+1)}{3k_{\mathrm{B}}T}. \tag{23}$$

This effect is illustrated in Fig. 2a, which presents the typical shape of powder pattern corresponding to the dipolar interaction, as well as surfaces of constant $\Delta\sigma$

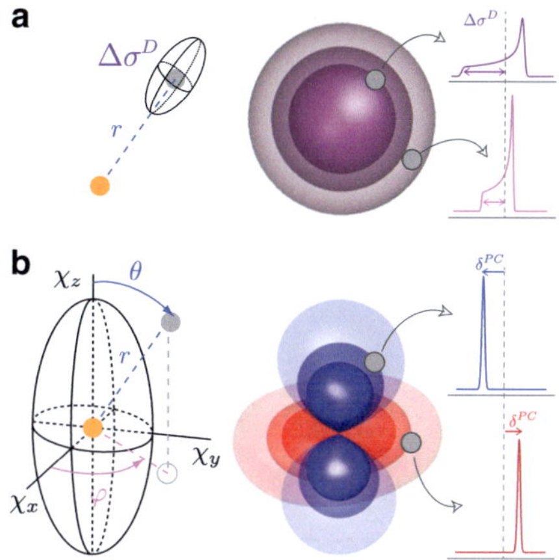

Fig. 2 Spatial dependence of the dipolar shift anisotropy (**a**) and of the pseudocontact shift in the magnetic susceptibility principal axis system (**b**). The position of the nucleus is defined by its spherical coordinates (r, θ, φ) in the PAS of the χ tensor. *Violet* surfaces represent isosurfaces of $\Delta\sigma$, and *blue* and *red* surfaces represent respectively positive and negative isosurfaces of δ^{PC}

around a metal center. Measurement of the shift anisotropy in a compound of known structure enables the susceptibility to be calculated. Alternatively, if the susceptibility is known, or can be estimated, it is possible to obtain information on the unpaired electron–nucleus distance.

Under MAS, if the angular spinning rate ω_r exceeds the magnitude of these anisotropic interactions, the same isotropic spectrum as obtained in the liquid state can be recovered. This corresponds to the rank-0 isotropic term of (22), the so-called PCS or δ^{PC}:

$$\delta^{PC} = W_{0,0} = \frac{1}{12\pi r^3} \times \left\{ \Delta\chi_{ax}(3\cos^2\theta - 1) + \frac{3}{2}\Delta\chi_{rh}\sin^2\theta\cos 2\varphi \right\}, \quad (24)$$

where $\Delta\chi_{ax}$ and $\Delta\chi_{rh}$ are the axial and rhombic components of the χ tensor, and the angles θ and φ describe the position of the nuclear spin with respect to the principal axes of the χ tensor, which depend solely on the geometry of the molecule, not on its orientation within the magnetic field. The dependence of the PCS on the geometry of the molecule is represented in Fig. 2b.

At spinning rates below the powder linewidth, that is, at rates which are not fast enough to average completely the second-rank paramagnetic interaction, the powder pattern observed in solids breaks up into sharp spinning sidebands. In this situation, beside the isotropic shift (the position of the "centerband"), the powder lineshape can be reconstructed from the relative intensities of the spinning sidebands. Since both PCS and shift anisotropies can be recovered, an increased amount of information is thus accessible in solid-state MAS NMR spectra.

Paramagnetic relaxation enhancements. The presence of a paramagnetic center in a molecule causes an increase in nuclear spin-lattice relaxation rates R_1 $\left(= T_1^{-1}\right)$ and spin-spin relaxation rates R_2 $\left(= T_2^{-1}\right)$ of the nearby nuclei, often referred to as paramagnetic relaxation enhancement (PRE).

The expression of these relaxation rate enhancements are given by the Solomon equations [24]:

$$R_1^{\mathrm{S}} = \frac{2}{15}\left(\frac{\mu_0}{4\pi}\right)^2 \frac{\hbar^2\gamma_{\mathrm{I}}^2 g_{\mathrm{e}}^2 \mu_{\mathrm{B}}^2 S(S+1)}{r^6} \times \left(\frac{\tau_{\mathrm{c}}}{1+(\omega_{\mathrm{I}}-\omega_{\mathrm{S}})^2\tau_{\mathrm{c}}^2} + \frac{3\tau_{\mathrm{c}}}{1+\omega_{\mathrm{I}}^2\tau_{\mathrm{c}}^2}\frac{6\tau_{\mathrm{c}}}{1+(\omega_{\mathrm{I}}+\omega_{\mathrm{S}})^2\tau_{\mathrm{c}}^2}\right), \quad (25)$$

$$R_2^{\mathrm{S}} = \frac{1}{15}\left(\frac{\mu_0}{4\pi}\right)^2 \frac{\hbar^2\gamma_{\mathrm{I}}^2 g_{\mathrm{e}}^2 \mu_{\mathrm{B}}^2 S(S+1)}{r^6} \times \left(4\tau_{\mathrm{c}} + \frac{\tau_{\mathrm{c}}}{1+(\omega_{\mathrm{I}}-\omega_{\mathrm{S}})^2\tau_{\mathrm{c}}^2} + \frac{3\tau_{\mathrm{c}}}{1+\omega_{\mathrm{I}}^2\tau_{\mathrm{c}}^2} + \frac{6\tau_{\mathrm{c}}}{1+\omega_{\mathrm{S}}^2\tau_{\mathrm{c}}^2} + \frac{6\tau_{\mathrm{c}}}{1+(\omega_{\mathrm{I}}+\omega_{\mathrm{S}})^2\tau_{\mathrm{c}}^2}\right). \quad (26)$$

In solids, the fluctuations of the magnetic field at the nucleus are mainly determined by electron relaxation ($\tau_{\mathrm{c}} = \tau_{\mathrm{e}}$), which occurs over timescales usually in the range 10^{-7}–10^{-13} s.

Note that in solution, the *average* electronic magnetic moment generates field fluctuations on the nucleus due to the stochastic reorientation of the molecule (the so-called Curie relaxation [25]). This relaxation mechanism is not operative in rigid solids [26].

3 Experimental Challenges

Direct extension of MAS NMR techniques to the routine analysis of paramagnetic organic and inorganic complexes is not straightforward, since the very same large orientation-dependent paramagnetic shifts, as well as paramagnetically-enhanced nuclear relaxation, hamper data acquisition and spectral assignment. The challenge is illustrated in Fig. 3, which shows the NMR spectra, acquired at 11 kHz MAS, of two microcrystalline samples, a diamagnetic Zn^{II} catalyst and its paramagnetic analog containing Fe^{II} [27].

In the diamagnetic case, suitable radio-frequency (RF) irradiation schemes allow the acquisition of well-resolved ^{1}H and ^{13}C spectra with linewidths almost comparable to their solution counterpart. The situation is completely different for the Fe^{II} complex, where very limited sensitivity and resolution are obtained by conventional MAS.

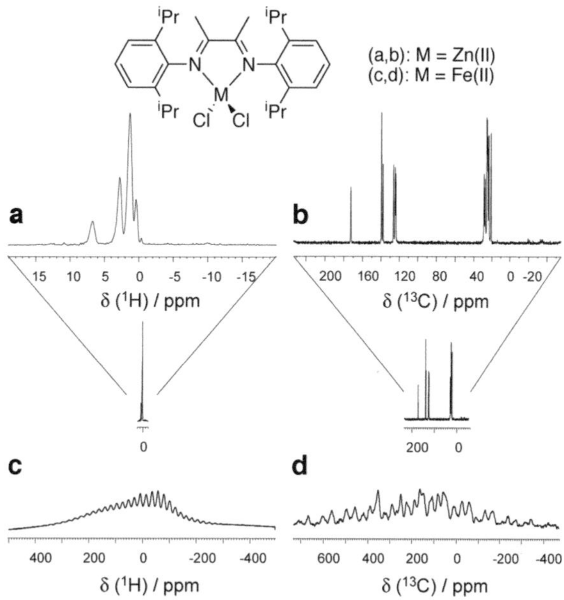

Fig. 3 An example of ^{1}H and ^{13}C MAS spectra of a diamagnetic Zn^{II} complex (**a**, **b**) and its paramagnetic Fe^{II} analog (**c**, **d**). Reprinted with permission from [27]. © 2006 American Chemical Society

3.1 Paramagnetic NMR Spectra and Fast MAS

A key observation for the solution of this problem was provided by Clayton et al. [23, 28], who showed that the major contribution to the line-broadening in paramagnetic organic compounds is not due to a direct effect of the unpaired electrons, but results from the inability to decouple the protons completely. The relatively large isotropic ^{1}H chemical shift dispersion and chemical shift anisotropies present in paramagnetic compounds make heteronuclear decoupling more demanding than in a typical diamagnetic case. Large RF bandwidths are needed for efficient heteronuclear decoupling and, moreover, the ^{1}H frequencies of the spins depend significantly on the orientation of the crystallite with respect to the magnetic field, making decoupling "on-resonance" difficult to achieve. These authors demonstrated that the ^{13}C MAS linewidth of lanthanide acetate tetrahydrate was substantially improved when both the acetate ion and the water of crystallization were replaced by deuterated forms (Fig. 4). Coupling between ^{13}C and ^{2}H and especially between ^{2}H and ^{2}H is weak enough for MAS alone to give line narrowing. A similar effect was reported for a deuterated organic free radical [23], where high resolution ^{2}H spectra could be obtained.

A similar problem occurs in the case of homonuclear decoupling. In highly deuterated materials, detection of resolved ^{1}H MAS NMR spectra was also shown to be possible. The linewidth of the residual protons is reduced by deuterium substitution, because the average ^{1}H–^{1}H distance increases and the lower magnetogyric ratio reduces the dipolar couplings. Nayeem and Yesinowsky reported the ^{1}H spectrum of $CuCl_2{\cdot}H_2O$, where the water of crystallization was replaced by the deuterated form.

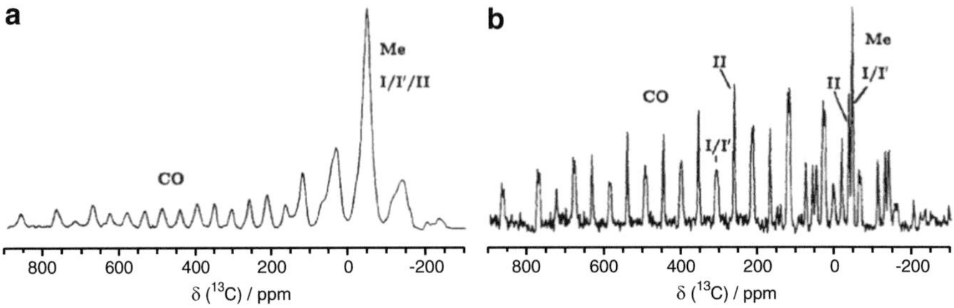

Fig. 4 ^{13}C MAS-NMR spectra of tetrahydrate europium acetate $Eu(O_2CCH_3)_3{\cdot}4H_2O$, in protonated (**a**) and deuterated (**b**) form, at a field of 4.7 T, at rotation rates of 4,650 Hz. Reprinted with permission from [23]. © 1993 American Chemical Society

Similarly, Brough et al. recorded a well-resolved ^{1}H spectrum of the residual protons in a largely deuterated sample of samarium acetate. However, problems of sensitivity and background suppression were encountered, since only a few percent of residual protonated groups could be afforded to obtain reasonable ^{1}H spectra.

In pioneering work, McDermott et al. [29] showed that, in favorable cases of paramagnetic centers with small (and isotropic) magnetic susceptibility (e.g., V^{II} and Cu^{II}), ^{1}H spectra at natural abundance could be resolved into separate lines at their respective isotropic chemical shifts and families of spinning sidebands (Fig. 5) simply by spinning at speeds of 7–15 kHz.

Correspondingly, when the proton spectra became resolved at rapid spinning speeds, it was observed that proton decoupling could become unnecessary for the acquisition of carbon spectra. In the case of $Cu(DL\text{-}Ala)_2$, while it was apparent that deuteration is more effective than proton irradiation in narrowing the carbon linewidths, it was also apparent that high-speed spinning is extremely helpful in narrowing the spectra and improving the sensitivity.

An explanation of the advantages offered by fast MAS lies in the fact that the large instantaneous paramagnetic shifts truncate the flip-flop interaction between protons, rendering the interaction inhomogeneous and allowing the fast MAS to refocus efficiently the homonuclear and heteronuclear couplings. This averaging is efficient regardless of large H resonance offsets and anisotropic shifts for paramagnetic interactions, while ^{1}H RF decoupling is more sensitive to these interactions in removing ^{1}H–^{13}C and ^{1}H–^{1}H dipolar couplings.

Recent technological developments in MAS probe design allowing spinning speeds up to the so-called "very fast" (40 kHz) and "ultra-fast" (70 kHz) MAS regimes have offered a powerful tool to extend the reach of these observations. First Ishii and coworkers [30, 31], and subsequently Pintacuda, Emsley and coworkers [26, 27] have shown that remarkable increases in resolution, in sensitivity, and in coherence lifetimes T_2' of the ^{13}C and ^{1}H spins are observed at spinning rates larger than about 25 kHz in the case of paramagnetic organometallic complexes (see Figs. 6 and 7). In these conditions, decoupling can be omitted without loss of resolution. This in turn drastically reduces the experimental duty cycle and allows one to utilize very short recycle delays afforded by rapid nuclear relaxation times in paramagnetic solids.

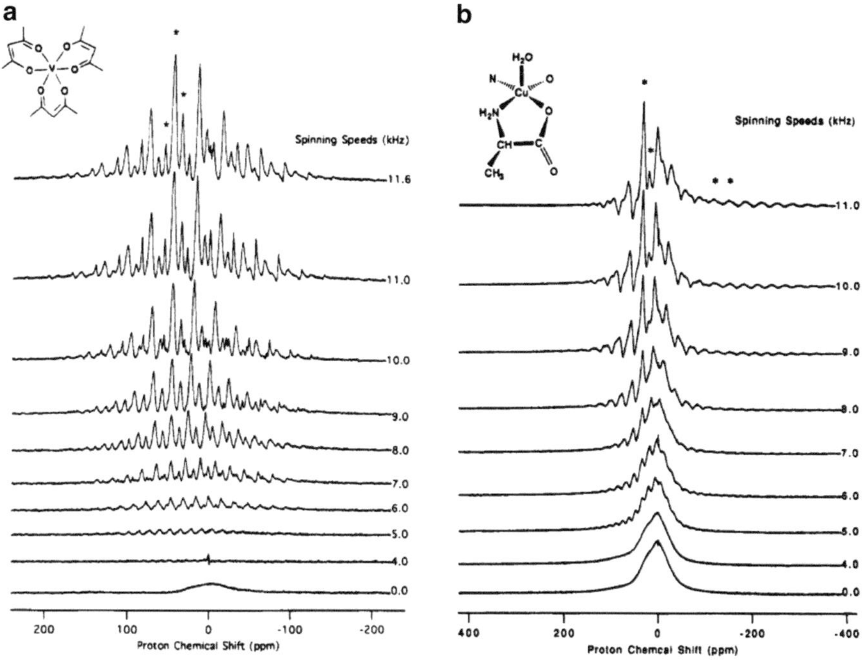

Fig. 5 A 400-MHz proton MAS spectra of V(acac)$_3$ (**a**) and of Cu(alanine)$_2$·H_2O. Full signal intensities were not achieved until the MAS spinning rate was above 11 Hz. Reprinted with permission from [29]. © 1995 American Chemical Society

These results are particularly striking if compared to the analogous behavior of a diamagnetic molecule, where, for example, the ^{1}H linewidth decreases linearly with the MAS speed [21], and the improvement in the linewidth obtainable by fast MAS alone is never sufficient to yield resolved spectra.

Hence, when sidebands are sufficiently suppressed by very fast MAS, the theoretical sensitivity of ^{1}H solid-state NMR for paramagnetic systems is greater than that for diamagnetic systems by an order of magnitude. Additionally, the spectra of paramagnetic species are better resolved by virtue of their broad chemical shift range. Sensitive and resolved spectra can thus be obtained within a few minutes for a small quantity (few milligrams) of samples at natural ^{1}H abundance in highly paramagnetic materials, allowing efficient detection of previously unobservable nuclei, and disclosing a rich amount of information that can be directly linked to the electronic and molecular structure.

3.2 Spin-Spin Correlations Under Fast MAS

Beside of the advantages connected with enhanced sensitivity and resolution obtained by using very fast spinning rates, very-fast MAS probes equipped with small RF coils permit one to apply high-power ^{1}H and ^{13}C fields covering large

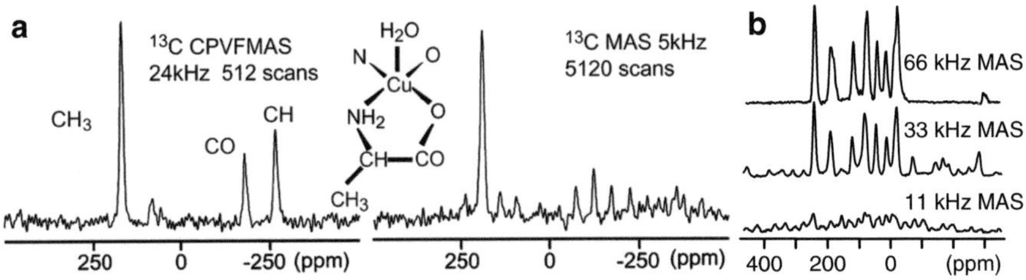

Fig. 6 (**a**) ^{13}C MAS spectra of Cu(DL-Ala)$_2$ at a field of 11.7 T, acquired at MAS rates of 24 kHz (*left*) and no ^{1}H decoupling, or at an MAS rate of 5 kHz, with 100 kHz ^{1}H CW decoupling. Reprinted with permission from [31]. © 2005 American Chemical Society. (**b**) Spinning speed dependence of ^{13}C MAS spectra of the Fe-DIAD ($S = 2$) complex at a field of 11.7 T, acquired without ^{1}H decoupling. Reprinted with permission from [27]. © 2006 American Chemical Society

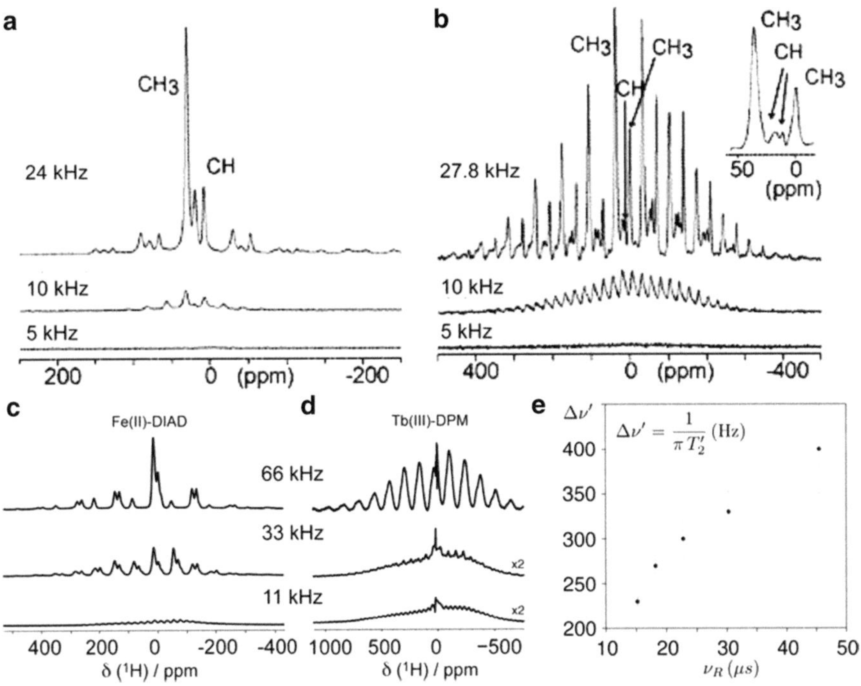

Fig. 7 Spinning speed dependence of ^{1}H MAS spectra of (**a–c**) Cu(DL-Ala)$_2$·(H_2O) and (**d–f**) Mn (acac)$_3$. The spectra were obtained at ^{1}H frequency of 400.2 MHz with 1-pulse excitation and a rotor synchronous echo. Reprinted with permission from [31]. © 2005 American Chemical Society. Spinning speed dependence of ^{1}H MAS spectra of (**a**) Fe-DIAD ($S = 2$) and (**b**) Tb (DPM)$_3$ ($J = 6$). The spectra were obtained at ^{1}H frequency of 500 MHz with 1-pulse excitation and a rotor synchronous echo. Reprinted with permission from [26, 27]. © 2006 and 2007 American Chemical Society. (**c**) Spinning speed dependence of the ^{1}H coherence lifetimes, measured with a CPMG sequence with adiabatic pulse trains (see Sect. 3.3)

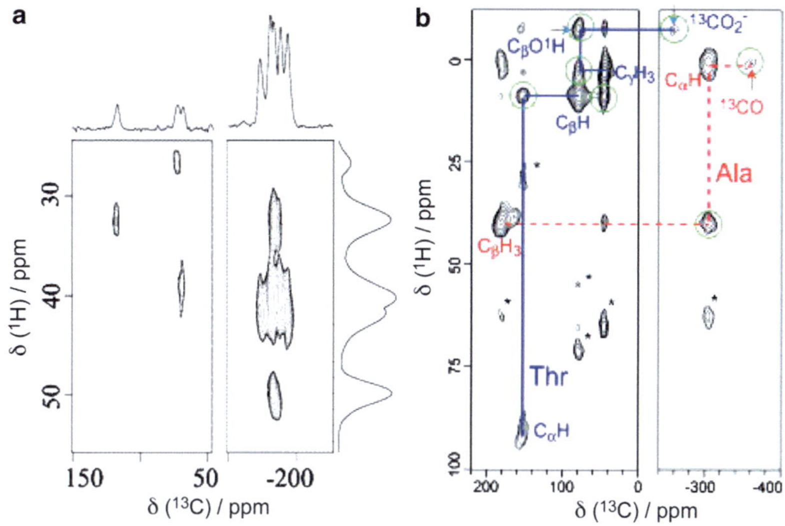

Fig. 8 2D $^{13}C/^{1}H$ CP correlation NMR spectra of (**a**) $V(acac)_3$, obtained at $\nu_R = 23$ kHz, and of (**b**) Cu(AlaThr) $\cdot$ $1/2H_2O$, obtained at $\nu_R = 25$ kHz, at a field of 11.7 T. Sample quantities were about 15 mg in the two cases, and experimental times were respectively 2.1 h and 6.1 h. Reprinted with permission from [30] (© 2003 American Chemical Society) and from [32] with permission from Elsevier

bandwidths. This opens the way to the recording of rapid 2D heteronuclear correlation and dipolar-edited experiments, which could be used to assign ^{13}C and ^{1}H signals, at least in simple model compounds [30, 32].

In this regard, it must be noted that there are several advantages offered by paramagnetic NMR in the solid state over solution NMR. Whereas in solution longer delays are typically required for obtaining through-bond transfers via the (smaller) J-couplings, dipolar-based polarization transfers in solids are efficient over contact times typically on the order of a few milliseconds, and interfere much less with the fast relaxation timescale of the spin systems. Rather, short proton T_1 values are turned into an advantage, allowing increased repetition rates, which are here only limited by the duty factor of the probe. Combined with the absence of decoupling fields, 2D spectra such as those displayed in Fig. 8 can be acquired.

Most importantly, solid-state signals are not subject to Curie broadening [26], which is the limiting factor in the study of large paramagnetic molecules by liquid-state NMR. This is of particular importance for the study of systems containing paramagnetic centers for which the correlation times for the electronic fluctuations (τ_e, see (25) and (26)) are short ($<10^{-11}$ s).

In solids, these centers (typically, Co^{II}, Fe^{III}, Ln^{III}) enhance significantly the longitudinal relaxation of the surrounding spins, but have a negligible effect both on their observed linewidths and on the coherence lifetimes, and do not significantly reduce the efficiency of magnetization transfer. In these conditions, dipolar-based experiments optimized for the fast MAS regime (double quantum CP [33, 34] or

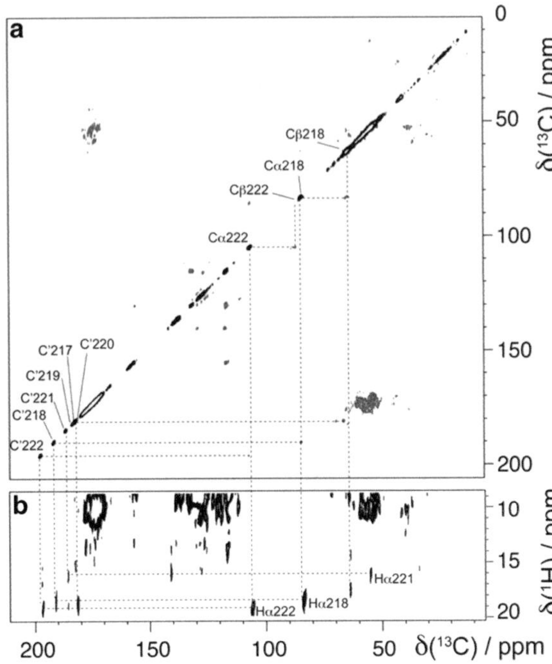

Fig. 9 Regions of DREAM (**a**) and HETCOR (**b**) correlations on microcrystalline (^{15}N, ^{13}C)-labeled CoMMP-12 recorded at 60 kHz MAS at 21.2 T. Reprinted with permission from [36], © 2010 American Chemical Society

DREAM [35]) provide efficient ways to generate correlations between these nuclear spins, similar to diamagnetic materials [36] (Fig. 9).

For highly paramagnetic materials, however, short ^{1}H longitudinal relaxation times (<1 ms) can impose strict limits even to dipolar-based transfers, and routinely available RF fields are unable to lock spins over the large bandwidths swept by the anisotropic shifts during a rotor cycle. For example, in the case of the Fe^{II} compound of Fig. 10, the CP experiment performs much worse than a ^{13}C direct excitation experiment under fast MAS. Transfer sequences that do not suffer from these drawbacks were developed to counter this kind of problem. In two parallel studies [27, 32] it was demonstrated that the use of a simple heteronuclear recoupling sequence like TEDOR (transferred-echo, double resonance) provides a broadband, offset-insensitive route to record much more efficiently 2D correlations on strongly paramagnetic signals. The pulsed nature of the sequence allows the use of higher ^{1}H and ^{13}C fields, and only one or two rotor cycles for the transfer period at high spinning speeds ensure a uniform transfer with optimal efficiency.

In these approaches, experimental support for resonance assignment can be provided by ^{1}H–^{13}C dipolar editing experiments [27, 30, 32, 37] in which the carbon spectrum is acquired after a REDOR-type recoupling sequence. ^{1}H–^{13}C dipolar couplings are switched on or off by 180° pulses applied on the ^{1}H channel and carbon signals experience differential dephasing depending on the number and the dynamics of the coupled protons (Fig. 11).

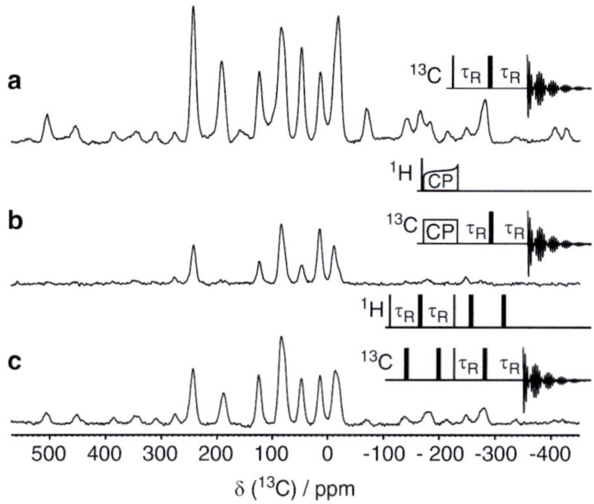

Fig. 10 Comparison between ^{13}C MAS NMR spectra of Fe-DIAD, obtained through direct carbon excitation using the sequence shown *inset* (**a**), via CP (**b**; contact time: 500 μs) and via the modified TEDOR sequence (**c**; recoupling time: 60 μs). *Narrow* and *wide rectangles* denote 90° and 180° pulses, respectively; τ_R represents the rotor period. Reprinted with permission from [27], © 2006 American Chemical Society

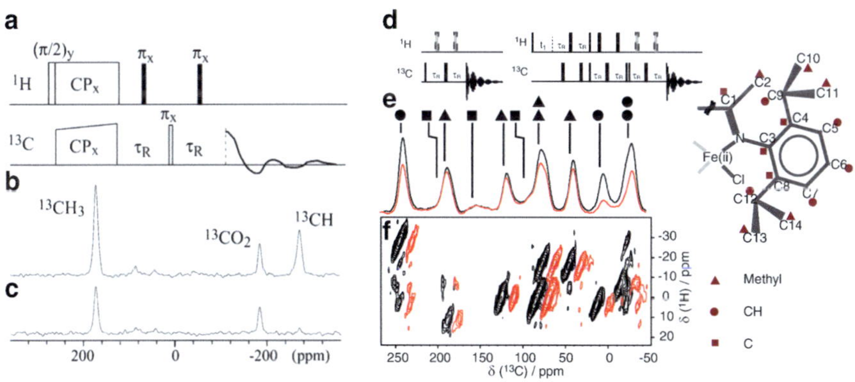

Fig. 11 (**a**, **d**) Pulse sequences for ^{13}C–^{1}H dipolar filter signal editing, where τ_R denotes one rotation cycle. Dipolar dephasing is introduced when the ^{1}H π-pulses (*filled boxes* in panel (**a**), *shaded boxes* in panel (**d**)) are applied. (**b**, **c**) ^{13}C CPMAS spectra of Cu(DL-Ala)$_2$·(H$_2$O) obtained at $\nu_R = 23$ kHz without (**b**) and with (**c**) ^{13}C–^{1}H dipolar dephasing (reprinted with permission from [30], © 2003 American Chemical Society). (**e**) 1D ^{13}C NMR and (**f**) 2D ^{1}H, ^{13}C-TEDOR NMR spectra of Fe–DIAD, obtained at $\nu_R = 33$ kHz, acquired without (*colored lines*) and with (*gray lines*) ^{1}H dipolar filter editing. Signals are labeled according to the correspondent carbon type: quaternary (*squares*), methine (*circles*), and methyl (*triangles*) (reprinted with permission from [27], © 2006 American Chemical Society)

3.3 Adiabatically Swept Pulses

Depending on the nature of the paramagnetic center, the shifts and shift anisotropies induced on the surrounding nuclei can be extremely large, reaching values that are larger than the practicable RF amplitudes. In these cases, the key barrier to further progress in the studies of several classes of new substrates is represented by the difficulty to excite and invert signals.

One possible solution to achieving broadband inversion is to use swept-frequency adiabatic pulses, which give very high bandwidths in relation to the RF power used. These pulses are widely used in solution state experiments, for instance for facilitating heteronuclear decoupling, or in many applications in magnetic resonance imaging (MRI) [38].

The treatment of adiabatic pulses in rotating solids is not straightforward, as the picture is deeply perturbed by the shift modulation imposed by the sample rotation. In recent years, we have introduced a framework for representing the Hamiltonian of the spin system in an NMR experiment under MAS, allowing one to define the criteria of phase and amplitude modulation for designing an adiabatic inversion pulse in rotating solids. This has led to the design of two families of pulses able to achieve efficient inversion of an entire sideband family of several hundred kilohertz, namely (a) very short adiabatic pulses (SHAPs) endowed with a wide and fast frequency sweep, whose behavior remains within the limit of the adiabaticity condition [39] and (b) slow, low-power adiabatic sweeps that irradiate a single sideband out of a broad anisotropic pattern but still result in complete inversion of the entire sideband pattern [40]. This latter strategy, similar to the approach used in the NMR of quadrupolar nuclei under MAS [41–44], produces efficient inversion of entire sideband families of several hundred kilohertz using low-power (10–60 kHz), single-sideband-selective adiabatic pulses, showing that the ratio between the RF power and the achieved bandwidth under MAS can be even larger than the solution-state counterpart (Fig. 12).

Once suitable pulses for manipulating largely anisotropic spin signals become available, these elements can be combined as building block for more complex pulse sequences. For example, adiabatic pulses greatly improve rotor-synchronous echo sequences, which are an important block for the acquisition of signals for patterns embracing very large spectral widths, as the observation of the second rotational echo removes significant baseline distortion and difficulties in phase-correction [27, 39, 40].

SHAPs were also employed to improve TEDOR polarization transfer in paramagnetic molecular complexes of iron, ytterbium and terbium, with shift anisotropies up to 1,200 ppm [39, 45]. Finally, SHAPs were employed as broadband adiabatic refocusing elements into a CPMG scheme, as an easy method to access paramagnetic linewidths in microcrystalline samples under MAS. This experiment allows one to concentrate the overall spectral intensity, initially distributed over the full dipolar anisotropy spinning sideband patterns, into a reduced number of spin-echo sidebands, and leads to a considerable increase in spectral sensitivity relative to experiments relying on MAS alone [26].

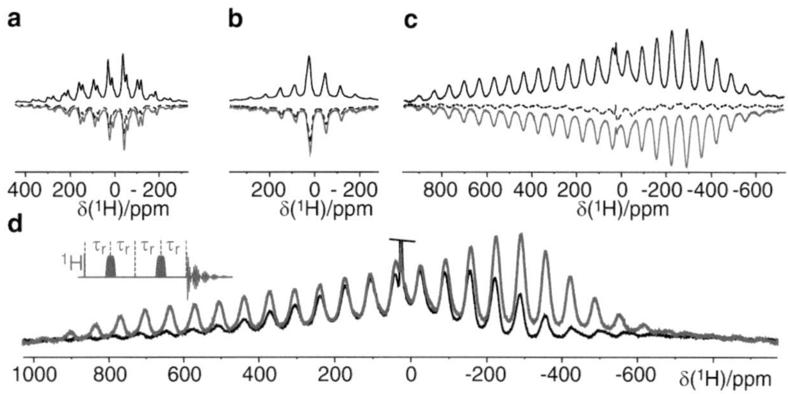

Fig. 12 ^{1}H spectra of Fe–DIAD ($S = 2$ (**a**)), Yb(DPM)$_3$ ($J = 7/2$ (**b**)), and Tb(DPM)$_3$ ($J = 6$, (**c**)), and their inversion recovery equivalents in two different cases: hard (*dashed line*) or adiabatic (*grey line*) inversion pulse. (**d**) Tb(DPM)$_3$ acquired under direct acquisition after a rotor synchronized double echo refocusing using SHAP (*gray*) or hard pulses (*black*). Reprinted from [39] with permission from Elsevier

4 Paramagnetic Probes in High-Resolution Solid-State MAS NMR

The efficient detection of high-resolution NMR signals in highly paramagnetic molecules (as outlined above) has opened news avenues toward the quantitative measurement of paramagnetic effects and their use for the structural characterization of more and more complex systems in different areas of chemistry, materials science, and biology.

In the following we review the use and the potential contained in the most striking changes produced in an MAS NMR spectrum by a paramagnetic center, notably in the shift, the shift anisotropy and the relaxation times of the surrounding nuclei.

4.1 Paramagnetic Shifts in Solid-State MAS NMR

Fermi contact shifts. A first area where paramagnetic solid-state NMR has played a unique role is that of *magnetic materials*, and in particular the chemistry-oriented area of molecule-derived magnetic clusters. The macroscopic magnetic properties of these objects are the result of interactions between the spin sources of the molecular building blocks. In the synthesis of new magnetic materials, particular attention is devoted to the search for strong intermolecular interactions, a prerequisite for a magnetic ordering of the spins [46].

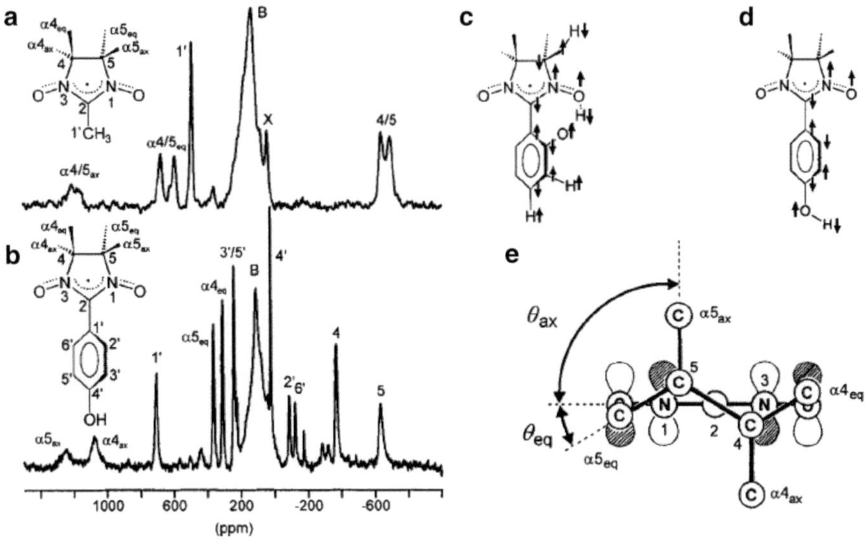

Fig. 13 (**a, b**) ^{13}C MAS NMR spectra of two nitronylnitroxide radicals (spinning rates of 10 kHz and 15 kHz, respectively), together with the corresponding sign patterns of the spin densities experimentally derived from the Fermi contact shifts (*up-spins* indicate positive spin). (**e**) Side view of an unsubstituted nitronylnitroxide down from the center of the bond C4–C5 to C2. The spin-carrying MO is located on the NO groups. Reprinted with permission from [51], © 1999 American Chemical Society

The most versatile and successful building blocks are transition metal complexes (e.g., Cu^{II} [47, 48], Mn^{III}, and Fe^{III} [49, 50] complexes) and nitroxide radicals [51–53] (Fig. 13).

For paramagnetic centers endowed with low susceptibility anisotropy, the dipolar contribution to the isotropic shift is negligible, and the observed isotropic paramagnetic shift is uniquely determined by the Fermi contact interaction, which is related to unpaired electron density transferred to or induced in ligand orbitals with *s*-atomic orbital content and, therefore, gives information about these delocalization mechanisms. Contact shifts can be quantitatively interpreted via hybrid Hartree–Fock/density functional theory (HF-DFT) calculations, the accuracy of which has nowadays significantly progressed [54, 55]. Direct conversion of NMR signal shifts into spin densities yield rather complete spin maps, from which in turn the spin delocalization mechanism can be deduced. In a manner analogous to that used for torsional angle dependence on ^3JCH or ^3JCC values, electron–nucleus hyperfine coupling constants can be related to the dihedral angle between the orbital containing the unpaired electron density and the carbon nucleus of interest. Moreover, while the experimental spin densities are in agreement with ab initio calculations on isolated molecules, deviations can be used to give insight into the mechanism of intermolecular interactions occurring in the solid phase. Most importantly, solid-state NMR spectroscopy allows for the measurement of paramagnetic shifts over a large range of temperatures, which is essential to reveal the

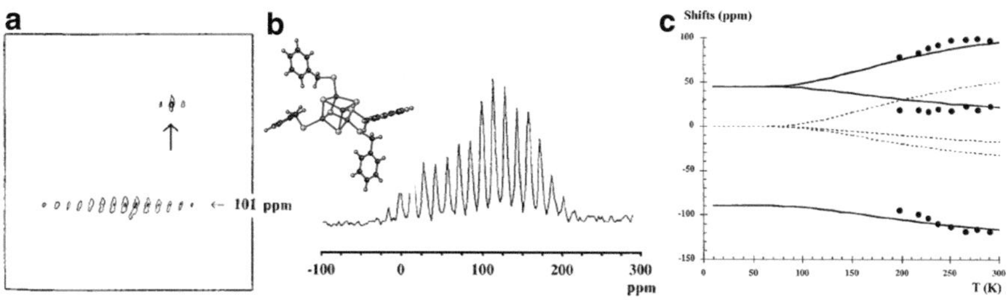

Fig. 14 Solid-state NMR spectroscopy gives access not only to the isotropic shifts but also to the principal values of the shift tensor (see Sect. 4.2). Emsley and coworkers [56] employed 2D ^{13}C magic angle turning correlation spectra (**a**, **b**) to record the temperature dependencies of both the isotropic and anisotropic parts (**c**) of the ^{13}C paramagnetic shift tensors in three $[4Fe–4S]^{2+}$ clusters, which are model compounds of the active sites of 4Fe–4S proteins. Strong antiferromagnetic exchange coupling constants among iron atoms could be deduced here. Reprinted with permission from [56], © 2000 American Chemical Society

mechanisms of the exchange interactions between the different paramagnetic centers [52, 53, 56, 57], and to derive accurately the spin coupling parameters within the cluster magnetic spin ladder. Notably, since NMR spectroscopy probes local magnetic effects, solid-state NMR determinations are not sensitive to traces of paramagnetic impurities as are bulk susceptibility measurements (Fig. 14).

Fast MAS, the use of low magnetic field strengths, and often the use of nuclei with low gyromagnetic ratios have also provided keys to many of the recent developments in the growing area of electrode materials for batteries, and in particular lithium-ion rechargeable batteries [58], where paramagnetic NMR can nowadays be used to determine the metal oxidation state and how it changes on battery cycling (critical to understand the redox processes that occur in these systems), and to elucidate the local environment [59] and the fate of the Li ions (directly involved in battery function) [60] (Fig 15).

Finally, access to unpaired electron density transferred to or induced in ligand orbitals is of particular interest in the area of *chemical catalysis*, where monitoring the electron spin states in a given metal center is a key step for understanding and controlling a reaction mechanism. For example, paramagnetic ^{1}H and ^{13}C shifts under fast MAS were employed to probe the fine details of the electronic state (frontier orbitals) in a high-spin ($S = 2$) Fe^{II} polymerization catalyst (Fig. 16 [27]), and similarly paramagnetic ^{15}N shifts were used to highlight reactive sites in oxidized flavin [61].

The solid-state NMR analysis of paramagnetic powders features in particular strategic applications for the understanding of solid-state reactions, and can strongly contribute to the development of *nano-scale and surface chemistry*. The ability to evaluate hyperfine shifts in paramagnetic complexes can help explain how the high reactivity of some systems and the lack of expected reactivity of others can be tuned by the choice of the solid support.

The seminal example in this area is represented by the work of Köhler and coworkers [62–64]. Paramagnetic solid-state NMR spectroscopy, along with detailed quantum mechanical calculations performed with different DFT functionals, has led

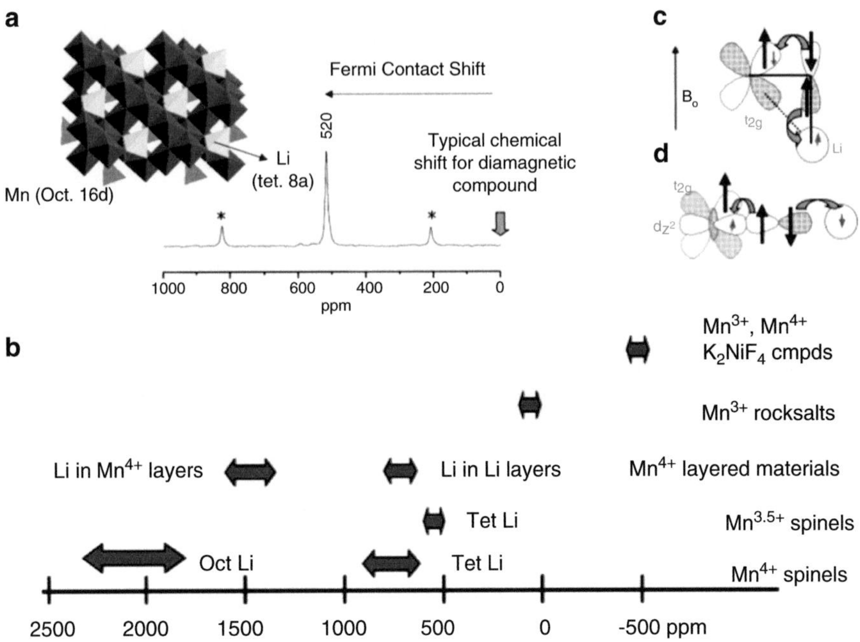

Fig. 15 ^{6}Li NMR shift is an extremely sensitive tool for the characterization of the local structures and the electronic properties of lithium manganese oxides, among the most common cathode materials in lithium rechargeable batteries (**a**). The major shift contribution in the ^{6}Li NMR spectrum arises from the hyperfine shift due to manganese ions in the first cation coordination sphere, so different shift ranges report on different lithium local environment (**b**). Moreover, these authors examined the local environments around lithium in a series of Mn^{IV} and Mn^{III} compounds, and rationalized the causes of the shifts in terms of both the nature and extent of the overlap between the manganese, oxygen and lithium orbitals (**c**, **d**). Reprinted from [60] with permission from Elsevier

to the assessment of the nature of surface active species in the Union Carbide catalyst, that can be synthesized by reaction of chromocene (Cp_2Cr) with silica (SiO_2), and allows the polymerization of olefins at low pressure and low temperature without any cocatalyst. More recently, paramagnetic solid-state NMR was used to follow the reactivity of supported Cr^{II} centers with CO, which is an established platform for assessing the efficiency of chromium-based systems in the field of olefin polymerization. Notably, solid-state NMR has allowed one to follow the spin flip reaction between Cp_2Cr (triplet spin state) and $Cp_2Cr(CO)$ (singlet spin state) on a polystyrene matrix at very low Cr percentages of the total active nuclei in the sample [65].

In a different application, the measurement of ^{13}C hyperfine shifts under very fast MAS was employed in combination with ab initio DFT calculations to characterize solid-state reactions involving rearrangements of polymorphs in a Cu^{II} paramagnetic drug [66] (Fig 17).

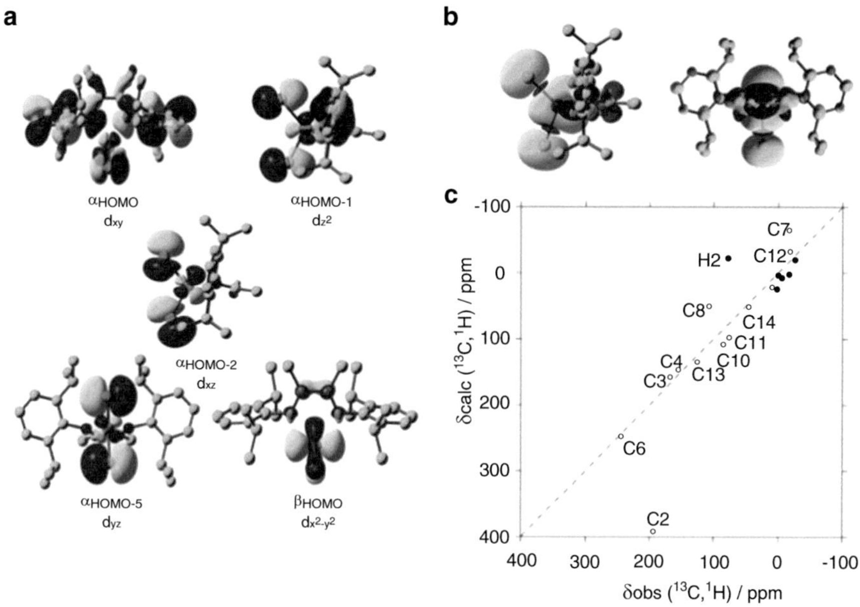

Fig. 16 (**a**) Isosurface representation of the frontier molecular orbitals for Fe-DIAD (DIAD: 2,3-dimethyl-1,4-[(2',6')-di-isopropylphenyl]-*N*,*N'*-diazadiene), and (**b**) spin density, superimposed on views of the molecular structure. (**c**) Comparison between the calculated and experimental shifts for the complex. Reprinted with permission from [27], © 2006 American Chemical Society

Pseudocontact shifts. As mentioned in Sect. 2, non-negligible zero-field splitting or strong spin-orbit coupling complicates the interpretation of the NMR effects caused by metal ions such as $Fe^{II/III}$, Co^{II}, and Ln^{III}. In these cases, however, the effect can be interpreted as a PCS, which is the result of the coupling between point-dipoles associated with the nuclear spins and the anisotropic magnetic susceptibility tensor. At intermediate distances from the paramagnetic centre where the spin density is not sufficiently small, both contact and "non-contact" mechanisms are operative, and a precise, quantitative separation of the observed shifts into different contributions is somewhat problematic.

A cumbersome but effective way of separating contact and pseudocontact contributions was proposed for lanthanides. Instead of analyzing the shifts for a single compound, this approach consists in considering the shifts in a series of homologous, isostructural compounds, under the assumption that the hyperfine constant ($\mathcal{A}$, at the origin of the contact shift) as well as the crystal field parameters (proportional to the $\Delta\chi$ tensor anisotropy, according to Bleaney's theory of magnetic susceptibility for lanthanide ions [67]) are approximately the same for all the compounds in the series. Contact and PCS should then vary regularly with trends corresponding respectively to that of the expectation value of the spin magnetization $\langle S_z \rangle_J$ (contact) and to that of a particular numerical function of the total magnetic moment J (pseudocontact), which are known and tabulated for the free

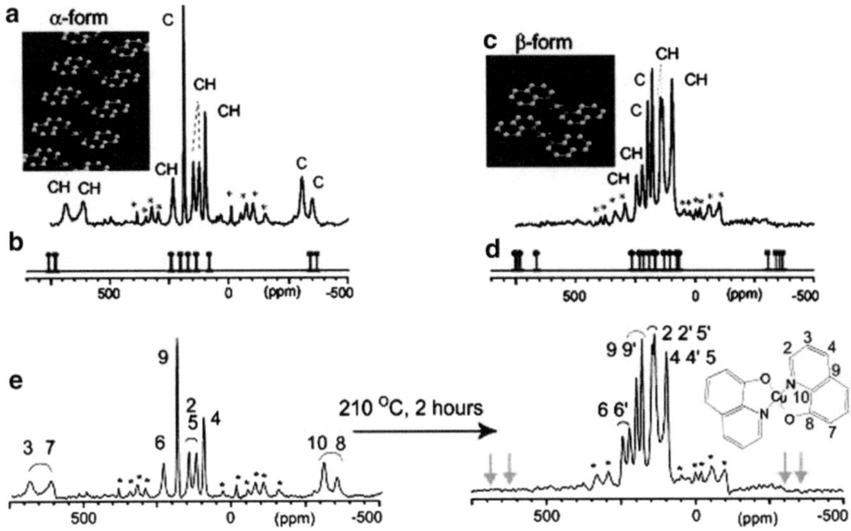

Fig. 17 (**a**, **b**) ^{13}C MAS spectra of (**a**) α-form and (**b**) *β*-form Cu(8-quinolinol)$_2$ (CuQ$_2$) obtained at ^{13}C frequency of 100.6 MHz without decoupling under very-fast MAS at 20 kHz, with the corresponding X-ray structures. (**c**, **d**) Shift positions obtained by ab initio calculations for (**c**) α-CuQ$_2$ and (**d**) *β*-CuQ$_2$ in the (*top*) monomeric and (*bottom*) tetrameric forms. (**e**) ^{13}C MAS spectra of α-CuQ$_2$ (*left*) and *β*-CuQ$_2$ (*right*) obtained by heating α-CuQ$_2$ at 210°C for 2 h, with ab initio based assignments. Reprinted with permission from [66], © 2007 American Chemical Society

metal ions (see Fig. 18a). This method was tested in the case of the pyrochlores $Ln_2Sn_2O_7$ [15]. Plots of the experimental ^{119}Sn (see Fig. 18b, c) and ^{89}Y shifts against the theoretical values of the contact and PCS suggest that the dominant contribution to the paramagnetic shifts is from the contact mechanism for ^{119}Sn, implying a small covalent interaction between the lanthanide ion and the ^{119}Sn atom. In contrast, in the same series of studies, the shifts were found to be predominantly caused by a dipolar interaction for ^{89}Y [16], consistent with the more ionic nature of Y^{III} compared to Sn^{IV}.

The pseudocontact contribution to the shift becomes increasingly important relative to the contact contribution, as the number of bonds separating the paramagnetic ion and the resonating nucleus increases. In the above work on solid pyrochlore solutions, a pseudocontact mechanism was inferred in the case of stronger paramagnets (e.g., Pr^{III} or Nd^{III}), where the ^{119}Sn nuclei were found to be sensitive not only to neighboring paramagnetic ions but also to paramagnetic ions in the second and third coordination spheres. The χ tensor anisotropy and its orientation were modeled on the basis of the site symmetry derived from the crystal structure, and the PCS were used to obtain detailed geometric information about the relative position of the NMR nucleus and the lanthanide ion [14, 15].

The long-range nature of the electron–nucleus coupling in systems with anisotropic susceptibility has produced a growing interest in the field of biomolecular NMR. The protein backbone and sidechains offer a large pool of nuclei far from the

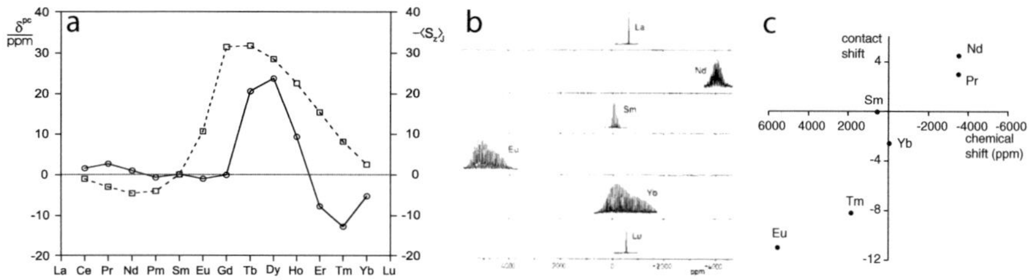

Fig. 18 (**a**) Predicted pseudocontact (*continuous line*) and contact (*dashed line*) lanthanide induced shift in a series of homologous compounds. For the contact shifts, the expectation values of the spin magnetization $-\langle S_z \rangle_J$ are those reported by Golding and Halton [68]; for the pseudocontact shifts, the values are those calculated by Bleaney [67]. (**b**) ^{119}Sn MAS NMR spectra of the pyrochlores $Ln_2Sn_2O_7$ at spinning speeds of 3–4 kHz, where Ln=La, Nd, Sm, Eu, Yb, and Lu. (**c**) Plot of the theoretically predicted magnitude of the contact shift vs. the experimental isotropic chemical shift for the lanthanide stannates (corrected for the diamagnetic shift of $Ln_2Sn_2O_7$). Reprinted with permission from [15], © 1989 American Chemical Society

metal center, free from any significant contact interaction, from which the Cartesian coordinate system defined by the χ tensor of a paramagnetic center can be identified. This coordinate system in turn presents a reference frame with respect to which surrounding nuclear spins can accurately be positioned using the geometric dependence of the PCS (25).

The role of PCS in this area was first demonstrated by McDermott and coworkers [69, 70] who used PCS in the solid state to contribute to an extension of the structure-activity relationship ("SAR by NMR") studies of ligand binding in a large paramagnetic protein. Cytochrome P450 BM-3 was studied using selective labeling schemes and relaxation filters to obtain site-specific assignments for some peaks, and a structural change involving a biochemically important amino acid residue (Phe87) upon ligation of the protein substrate (*N*-palmitoylglycine) could be monitored by changes in the PCS induced by the high-spin Fe^{III} center. Notably, in a microcrystalline samples of a metalloprotein, the metal centers are diluted by the large molecular size, thus limiting the anisotropic bulk magnetic susceptibility (ABMS) broadenings [22, 71, 72] occurring when paramagnetic metal ions are inside a continuous solid or in a micro-crystalline powder of highly-packed, small molecules.

More recently, PCS have been shown to possess a unique potential in the structural characterization of fully isotopically enriched protein systems at higher resolution. Bertini and coworkers [73, 74] showed that PCS generated by paramagnetic metal ions with sufficiently large spin quantum number and magnetic susceptibility anisotropy can be easily measured, in large numbers, in solid-state spectra of microcrystalline proteins. About 250 PCS were observed in microcrystalline Co^{II}-substituted matrix metalloproteinase 12 for nuclei up to more than 20 Å from the metal, and found in very good agreement with the calculated ones. The structural information contained in these PCS allows one to reduce greatly the number of ^{13}C–^{13}C distance restraints needed to obtain a protein structure. These findings open new perspectives for solid-state protein structure determinations in solid-state

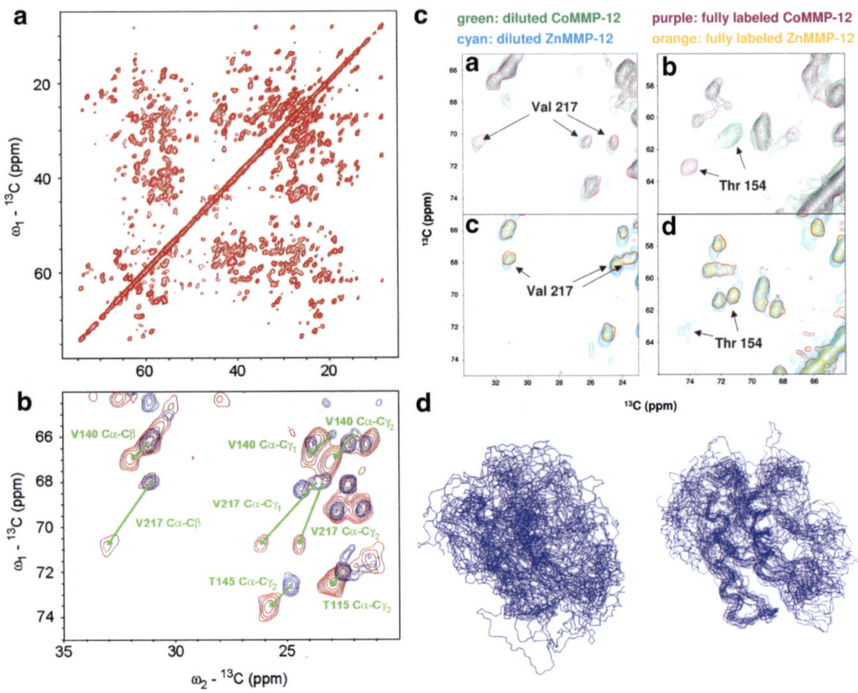

Fig. 19 (**a**) ^{13}C–^{13}C CP-MAS PDSD spectrum of a micro-crystalline sample of Co^{II}-substituted MMP-12 (16.4 T, $\nu_R = 11.5$ kHz, mixing time 60 ms, 290 K). (**b**) Superposition of the ^{13}C–^{13}C CP-MAS PDSD spectra of the diamagnetic zinc MMP-12 (*blue*, 16.4 T, $\nu_R = 11.5$ kHz, mixing time 15 ms) with the paramagnetic CoMMP-12 (*red*). *Green arrows* indicate the paramagnetic shifts. (**c**) Regions from PDSD spectra of fully labeled ZnMMP-12 (*orange*), fully labeled CoMMP-12 (*purple*), diluted CoMMP-12 (*green*), and diluted ZnMMP-12 (*cyan*). (*Sub-panels* A and C) The peaks of Val-217, which is affected only by intramolecular pcs, and thus the shifts observed in full-labeled samples are analogous to those observed in diluted samples. (*Sub-panels* B and D) The peaks of Thr-154, which is strongly affected only by intermolecular pcs, and the shifts observed in fully-labeled samples differ from those observed in diluted samples by the intermolecular contributions. (**d**) Families of 15 structures obtained without paramagnetic restraints (*left*) and with paramagnetic restraints (*right*). Readapted with permission from [73] (© 2007 American Chemical Society) and [75] (© 2008 National Academy of Sciences, USA)

NMR, where the determination of a dense network of long-range restraints between carbon spins constitutes an ongoing challenge [75] (Fig 19).

In a crystal, however, the observed shifts may be a complicated superposition of inter- and intramolecular dipolar interactions. Bertini and coworkers [75] showed how a combined strategy of protein labeling and dilution of the paramagnetic species allows one to separate easily the PCS contributions originated by the protein internal metal (intra-molecular PCS) from those due to the metals in neighboring proteins in the crystal lattice (inter-molecular PCS). The inter-molecular PCS provide unique information on the protein arrangement in the solid phase, determining positions and orientations of neighboring proteins, in a sort of “NMR crystallography". Along the same lines, Kervern et al. showed that, in the case of a series of small lanthanide complexes, inter- and intramolecular PCS measured from ^{1}H spectra can be simultaneously employed for a full crystal structural determination, with no prior knowledge on the metal χ tensor anisotropy, if the anisotropy of the PCS is measured at the same time together with their isotropic values [45] (see Sect. 4.2).

4.2 Shift Anisotropies in Solid-State MAS NMR

There have been several studies of shift anisotropy in paramagnetic solids, obtained from the orientation analysis of powder lineshapes. Bloembergen [76], Rundle [77], and Poulis and Hardeman [78] analyzed the orientation dependent ^{1}H shifts in single crystals of hydrated Cu^{II} salts at low temperatures. McGarvey and Nagy [79, 80] have investigated the temperature dependence of static powder ^{1}H spectra of uranocene $U(C_6H_6)_2$, a system where all the protons are chemically equivalent.

Single crystals are not, however, available for many compounds, and analysis of powder spectra is often difficult because of the need to deconvolute powder patterns from different resonances. More conveniently, MAS allows separation of resonances in spectra with overlapping powder patterns; the overall shift anisotropy information is then contained in the sidebands.

Dipolar shift anisotropy. The first quantitative interpretation of the intensity pattern of spinning sidebands observed in MAS NMR spectra of paramagnetic solids was provided by Nayeem and Yesinowski in an investigation of ^{1}H MAS spectra acquired from residual protons in a largely deuterated sample of polycrystalline $CuCl_2{\cdot}2H_2O$.

In a microcrystalline sample, the crystalline packing exposes each nucleus to the field generated by the paramagnetic centers of neighboring molecules in addition to intramolecular paramagnetic interactions. Contributions from intercomplex hyperfine couplings are not negligible, particularly for small complexes. Interestingly, however, these authors showed that, to a good approximation, the local fields on a nucleus are independent contributions from the individual electrons surrounding it [81]. The susceptibility can be estimated on the basis of an isotropic free ion model and, from the knowledge of the crystal structure, the total electron dipolar Hamiltonian can be easily obtained by summing up over the lattice the dipolar matrices corresponding to the individual electron–nucleus spin pairs. This calculation is relatively simple and based on the periodicity of the crystal lattice and the local symmetry of the molecule, although it proves necessary to sum over large distance ranges (e.g., up to 50 Å) for convergence.

Over the years, the interpretation of the sidebands observed for different nuclear spins has been employed to probe local structure: the central transition in the ^{23}Na MAS NMR spectra from $NaLn(EDTA){\cdot}8H_2O$ (Ln=Nd, Pr, Eu) [82], ^{13}C in Ln^{III} acetates [23] or in hydrated Cu^{II} salts [32], ^{6}Li and ^{7}Li in lithium manganates for

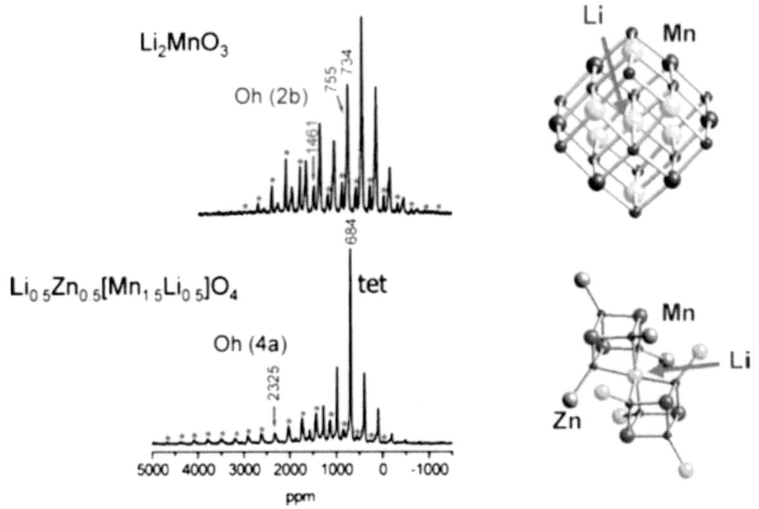

Fig. 20 The ^{6}Li MAS NMR spectra of the ordered spinel $[Li_{0.5}Zn_{0.5}]_{tet}[Li_{0.5}Mn_{1.5}]_{Oct}O_4$ and Li_2MnO_3. The *asterisks* indicate the spinning sidebands of the resonances from Li in the octahedral site in the spinel and the 2b site of Li_2MnO_3 (i.e., the site due to Li in the Mn layers). The local environments for Li in these sites are shown to the right of the spectra. Reprinted from [60] with permission from Elsevier

use as cathode materials in lithium rechargeable batteries [60, 83], ^{31}P in Ln^{III} substituted polyoxoanionic (Keggin and Wells-Dawson) solids [84, 85]. In all these cases, the computed paramagnetic dipolar tensors are in very good agreement with the experimental spectra, provided the dipolar couplings of the observed nucleus to surrounding nuclei, including protons, are sufficiently small, or can be eliminated, either by very-fast MAS, or by deuteration, so that the paramagnetic dipolar contribution to the shift anisotropy is dominant. In the latter case, ^{2}H itself constitutes a useful NMR probe, where the way the quadrupolar lineshape is modified by the paramagnetic dipolar shift anisotropy is a powerful reporter for both structure and dynamics [29, 86–91].

Because of the need to sum over many sites in the crystal, however, the approach is most valuable to test the validity of proposed models rather than to determine a priori structures. For example, Nayeem and Yesinowski [81] made use of this approach to estimate the degree to which paramagnetic electron density is delocalized from the copper ion onto the chloride ions. Several calculations were performed assuming variable degree of electron spin delocalization, until a closer agreement was reached to the observed spectral width, so as to reproduce the ^{1}H sideband pattern reasonably well. In a recent application, Grey and coworkers [60, 83] used the signs for the anisotropy of the ^{6}Li dipolar tensor to characterize different geometric arrangements of the Mn^{IV} ions around the Li ions in spinel lithium manganates (Fig. 20).

In a reverse approach, Ishii and coworkers [92] examined the possibility of using the ^{1}H shift anisotropies to determine the electron-spin states for Fe^{III} in an

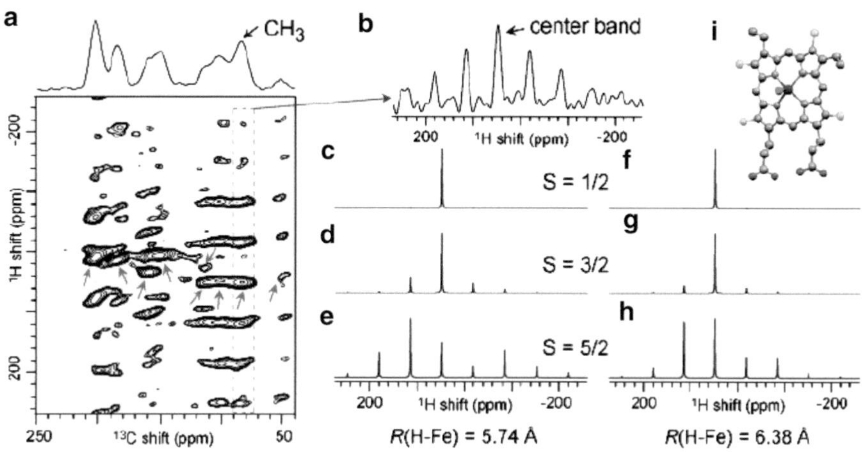

Fig. 21 (**a**) 2D $^{13}C/^{1}H$ correlation NMR spectrum of hemin chloride obtained with dipolar INEPT transfer at a spinning speed of 26.3 kHz, together with (**b**) a slice along the ^{1}H dimension of the CH_3 peak marked with a *red box*. (**c**–**h**) Corresponding simulated ^{1}H spinning sideband patterns for electron spin states of S = (**c**, **f**) 1/2, (**d**, **g**) 3/2, and (**e**, **h**) 5/2 from the Fe–H distances for (**c**–**e**) the lower limit of 5.74 Å and (**f**–**h**) the higher limit of 6.38 Å. Reprinted with permission from [92], © 2007 American Chemical Society

unlabeled hemin chloride (chloroprotoporphyrin-IX), for which the metal-binding structure is relatively well defined (Fig. 21).

Summing over many sites in the crystal results in nuclear shift tensors which deviate from axial symmetry. Figure 22b shows how the network of reference points in crystal lattice alters the shape of the isosurfaces of $\Delta\sigma$. This reduces the precision in the extraction of intracomplex distances from the shift anisotropies if the asymmetry η is not taken into account [37]. More accurate distance information can be obtained in the case of a dilute solid solution of paramagnetic ions in a diamagnetic host, where strong dipolar coupling of a nucleus to more than one paramagnet is unlikely.

Another source of inaccuracy for most of these nuclei is represented by the large, non-negligible contribution of the diamagnetic CSA, which may pollute the experimental measurements, yielding a combined interaction rather than the electron-nuclear dipolar anisotropy alone. Simulations where the CSAs are added to the hyperfine tensors indeed provide a better approximation to the observed spectra. However, as described in the case of lanthanide acetates by Brough et al. [23], this correction is experimentally demanding, as it requires the measurement of ^{13}C CSA tensors in a diamagnetic analog complex, and their alignment in the molecular frame using the orientations determined in single crystal NMR studies.

To overcome this problem, Kervern et al. explored the potential for structural investigation contained in the anisotropies of ^{1}H nuclei, for which CSA is negligible. ^{1}H shift anisotropy patterns in a class of lanthanide compounds were recorded at high-resolution in an adiabatic TEDOR experiment [45], and were shown to

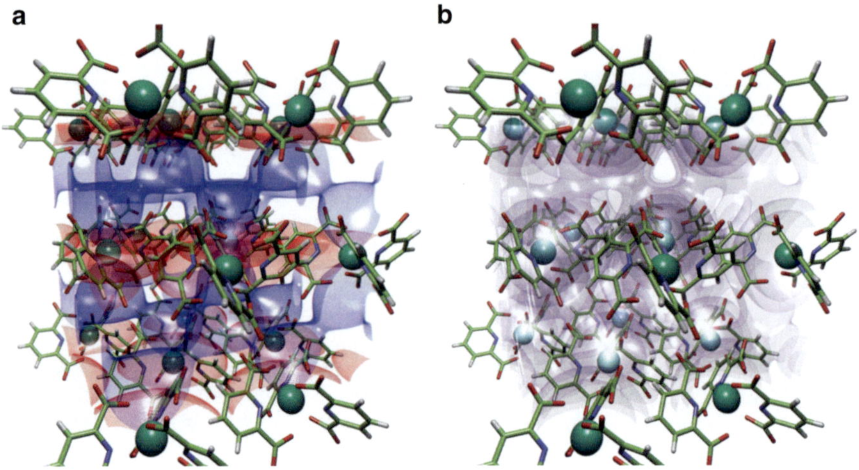

Fig. 22 Iso-surfaces for the pseudocontact shift (**a**) and the axial anisotropy of the chemical shift tensor (**b**) superimposed on the crystalline lattice of $Yb(DPM)_3$. The PCS isosurfaces constitute a representation of the χ tensor, while the $\Delta\sigma$ isosurfaces constitute a representation of χ_{iso}. Readapted with permission from [45], © 2009 Wiley

provide simultaneously not only the geometrical information contained in the shift anisotropy of each proton, but also the geometrical information contained in the PCS. The combination of the effects offered a set of observables that severely restrained the position of each nucleus around the metal centers, defining not only the conformation of the molecule in the lattice but also the intermolecular packing in the solid phase [45] (Fig. 23).

Contact shift anisotropy. A particular case of shift anisotropy is that stemming from the dipolar interaction between a nuclear spin and the spin density located outside the *s* core of a molecule. While the Fermi contribution coming from the electron spin density located directly at the nuclear site, that is, in the *s* orbital of the atom, is isotropic, the spin density located in the atomic *p, d* or *f* orbitals may produce a traceless anisotropic coupling to the nuclear spin.

This phenomenon can be described invoking a second-rank hyperfine coupling tensor $\mathcal{A}_{2,m}$ [93], of components:

$$\mathcal{A}_{2,0} = \alpha \frac{\mu_0}{4\pi} \hbar g_e \mu_B \gamma_I \left\langle \frac{1}{r_{orb}^3} \right\rangle \rho_{orb}, \tag{27}$$

$$\mathcal{A}_{2,\pm 1} = \mathcal{A}_{2,\pm 2} = 0, \tag{28}$$

where ρ_{orb} is the fraction of electron spin *S* located in the occupied *p*, *d*, or *f* orbital of the atom under consideration, and $\left\langle \frac{1}{r_{orb}^3} \right\rangle$ is the expectation value of $\frac{1}{r_{orb}^3}$ (r_{orb} is the nucleus-electron distance) over the orbital (for a 2*p* orbital of a carbon atom, $\alpha = \frac{2}{3}$ and $\left\langle \frac{1}{r_{orb}^3} \right\rangle \rho_p$ is 13.5 Å^{-3}).

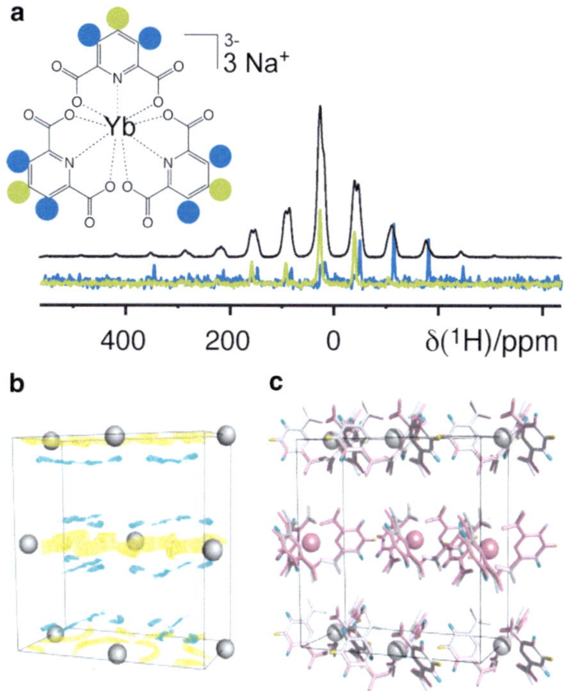

Fig. 23 (**a**) Slices of a TEDOR spectrum corresponding to each proton-bearing carbon site of Yb $(DPM)_3Cs_3$. (**b**) Agreement maps between experimental data and positions in the unit cell for the meta protons (*blue mesh*) and the para protons (*yellow mesh*) of Yb $(DPM)_3Cs_3$. (**c**) Structure of complex $Yb(DPM)_3Cs_3$ obtained by fitting the rigid dipicolinate ligands into the agreement maps. Readapted with permission from [45], © 2009 Wiley

This produces an anisotropy of the nuclear shift $\Delta\sigma^C$ equal to

$$\Delta\sigma^{\mathrm{C}} = -\frac{\chi\mathcal{A}_{2,0}}{\mu_0\mu_{\mathrm{B}}g_{\mathrm{e}}\hbar\gamma_{\mathrm{I}}}. \tag{29}$$

Note, however, that the $\mathcal{A}$ anisotropy, even though present, has the same transformation properties as the through-space dipolar coupling (described in the previous paragraph), and it is often difficult to distinguish from it. However, dipolar terms which also contribute to the total anisotropy of the signal shift can be estimated and factored out if the geometry of the molecule is known. In this case, a tensor analysis of the signals can give information about unpaired electrons outside the *s* core of the molecules. In two seminal studies [93, 94] (Fig. 24), Heise et al. determined the shift anisotropy from the spinning sideband manifolds of ^1H and ^{13}C spectra of a series of open-shell bis(cyclopentadienyl)metal compounds (metallocenes). The shift anisotropy was then converted to the anisotropic part of the hyperfine interaction tensor and subsequently to the spin densities, showing that NMR observables are a straightforward probe of the occupancy of the frontier orbitals of these sandwich compounds, which in turn tunes their intermolecular magnetic interactions. For example, in metallocenes with more than 18 valence electrons (i.e., cobaltocenes and nickelocenes), the spin density in the ligand π system is positive, and indeed the unpaired electrons occupy e^*_{1g} orbitals, which have considerable ligand contributions. In low-spin metallocenes with less than 18 valence

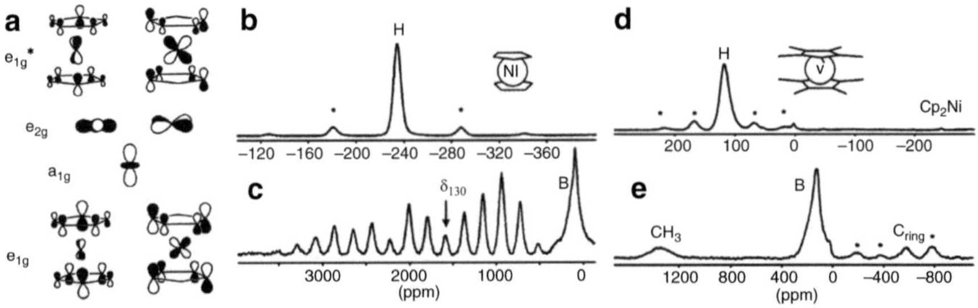

Fig. 24 (**a**) Frontier orbitals of metallocenes. (**b**, **c**) ^1H and ^{13}C MAS NMR spectra of Cp_2Ni (322 K, spinning rate 16.1 kHz). (**d**, **e**) ^1H and ^{13}C MAS NMR spectra of $(C_5Me_5)_2V$ (307 K, spinning rate 15.0 kHz). Reprinted from [93] with permission from Elsevier

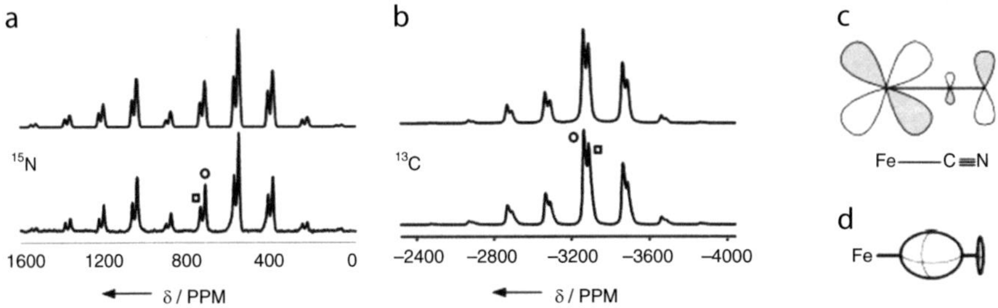

Fig. 25 (**a**) Experimental (*bottom*) and back-calculated (*top*) ^{13}C and ^{15}N MAS NMR spectra of $[Fe(CN)_6]^{3-}$ at 326.4 K, at MAS rates of 15 kHz and 5 kHz, respectively. The axial and equatorial isotropic signals are labeled with *squares* and *circles*, respectively. (**c**) *p*-Type orbital contributions of an M-CN fragment qualitatively representing the transfer of positive spin density from the metal to the CN ligand. (**d**) Sketch of the anisotropic spin distribution about the C and N atoms of the Fe–CN fragment as derived from NMR data. Reprinted with permission from [49], © 2004 Wiley

electrons (i.e., vanadocenes, chromocenes, and highly alkylated manganocenes), negative spin density is transferred to the ligand π system. This proves that the unpaired electrons are located in the metal-centered e_{2g} and a_{1g} orbitals, and instead, polarization is induced from the fully occupied bonding e_{1g} orbitals.

In a more recent contribution, Köhler and coworkers used ^{13}C and ^{15}N MAS NMR to determine quantitatively small spin densities on the CN ligands in $[Fe(CN)_6]^{3-}$ ($S = 1/2$) and $[Mn(CN)_6]^{3-}$ ($S = 1$), which are prominent building blocks of magnetic materials (Prussian-blue-type magnets) [49]. Interestingly, from the shift anisotropies, the spin distribution in different directions of the crystal lattice could be distinguished as well (oblate and prolate spheroids at C and N, respectively), indicating that the induction of spin at the C atoms involves polarization of electrons in *s*-type orbitals parallel to the M-CN axis, while direct spin transfer to the N atoms involves *p*-type orbitals perpendicular to that axis (see Fig. 25).

4.3 Paramagnetic Relaxation Enhancements in Solid-State MAS NMR

Enhanced longitudinal and transverse relaxation (PRE) of the nuclear spins surrounding unpaired electrons is often seen as a barrier to resolution and sensitivity solid-state NMR. At the same time, this phenomenon provides a number of advantages for structure determination that have been recognized early in the study of paramagnetic systems.

Paramagnetic doping is the most common approach for those systems where long relaxation times limit the sensitivity of the NMR spectra. For example, the use of low concentrations of iron and manganese to reduce the relaxation times of ^{29}Si nuclei in glasses and cements has long been a well-established practice [95].

This ability to observe the fast relaxing nuclei in local environments with very low concentrations becomes particularly interesting when paramagnetic ions are incorporated at low abundance within a solid. In particular, the resonances of different species close to the paramagnetic centers can be selectively observed by varying the recycle times between pulses and saturating selectively the resonances of nuclei far from the paramagnetic ions (Fig. 26) [14–16].

A new potential for paramagnetic doping has emerged in the field of biomolecular NMR, when coupled to new recent hardware developments, which have opened the way to the so-called ultra-fast (>60 kHz) MAS with small diameter rotors (e.g., 1.3 mm). These spinning rates enter a regime where spin dynamics is significantly altered and homonuclear ^{1}H–^{1}H couplings are efficiently averaged, removing the need for high RF irradiation fields and allowing recording of optimal ^{13}C and ^{15}N spectra employing RF fields which do not exceed 40 kHz during the whole experiment [33, 34, 96]. This is of particular advantage for the study of biological substrates with high dielectric constants, where prolonged high-power irradiations can lead to undesired, often disastrous, heating of the sample. With the aid of paramagnetic doping, the protein proton T_1s can be significantly shortened and, in combination with low power heteronuclear decoupling at very fast MAS, this enables the use of recycle delays as short as few tens of milliseconds, thus greatly reducing the experimental times for the acquisition of multidimensional correlations. This kind of sensitivity enhancement using paramagnetic ion doping has recently been proposed for ^{13}C solid state NMR of non-paramagnetic microcrystalline proteins, either in the presence of Cu^{II} doping [97, 98] or by incorporation of a thiol-specific EDTA-metal reagent (*N*-[*S*-(2-pyridylthio) cysteaminyl]EDTA) bound to Cu^{II} and Mn^{II} ions [99].

Relaxation data also represent a direct source of structural measurements, as the PRE depends on the distance between observed nuclei and the paramagnetic center. While fast spin-diffusion in solids complicates the quantitative measurement of ^{1}H longitudinal relaxation, significant variations in T_1 enhancements are indeed often observed for the different low-γ nuclear sites in paramagnetic compounds. However, these enhancements are generally the result of interaction of a nucleus with a complex network of paramagnetic sources in the crystal lattice, which makes their

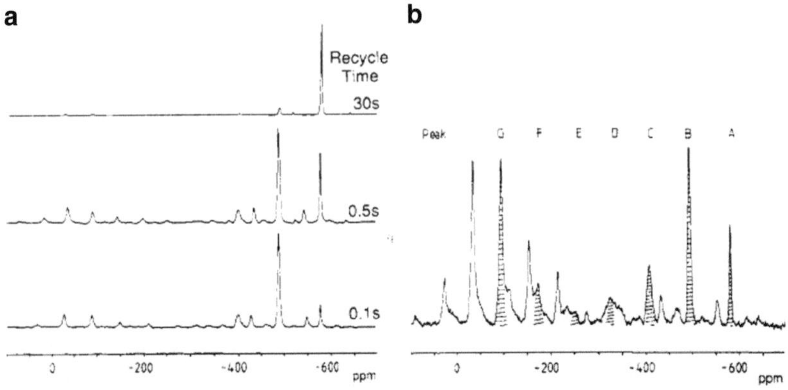

Fig. 26 (**a**) ^{119}Sn MAS NMR spectra from the preparation with stoichiometry $Y_{1.8}Sm_{0.2}Sn_2O_7$, obtained with recycle times of 30, 0.5, and 0.1 s. (**b**) ^{119}Sn MAS NMR spectrum of "$YSmSn_2O_7$" obtained with a recycle time of 0.5 s. The isotropic resonances A–G are *shaded*. Reprinted with permission from [15], © 1989 American Chemical Society

conversion into intracomplex distances difficult, or (when this is attempted) reduces the resulting precision [37]. Less ambitiously, more often, differential relaxation enhancements are only used to assist in the resonance assignment process on the basis of a structural model.

The possibility of exploiting enhanced paramagnetic relaxation as an effective source of structural constraints was explored by Jaroniec and coworkers on microcrystalline samples of a model protein, the B1 immunoglobulin-binding domain of protein G (GB1). Here a solvent-exposed cysteine residue was used to incorporate a thiol-specific paramagnetic nitroxide (TEMPO) or a thiol-specific EDTA-metal reagent bound to Cu^{II} and Mn^{II} ions [99–101].

Paramagnetic moieties exhibiting different relaxation times and spin quantum numbers modulate differently the longitudinal (R_1) and transverse (R_2) relaxation rates of the protein ^1H, ^{13}C, and ^{15}N nuclei. Specifically, nitroxide, Cu^{II} and Mn^{II} spin-labels cause considerable longitudinal nuclear PREs. The nitroxide and EDTA-Mn^{II} spin labels also generate large distance-dependent transverse relaxation enhancements, while this phenomenon is significantly attenuated for the Cu^{II} center.

The combination of negligible transverse and substantial longitudinal relaxation enhancements obtained with the EDTA-Cu^{II} side chain was shown to be especially advantageous to collect structural restraints for most sites in the protein. The measured longitudinal PREs for backbone amide ^{15}N nuclei were found to be highly correlated with the proximity of the Cu^{II} ion to ^{15}N spins, with significant effects observed for nuclei up to $\simeq 20$ Å away, thereby providing important information about protein structure on length scales that are inaccessible to traditional solid-state NMR techniques (see Fig. 27).

A different example of the use of paramagnetic relaxation for distance measurements was provided by the determination of the depth of insertion of the antimicrobial peptide protegrin-1 into lipid bilayer membranes [102]. By doping

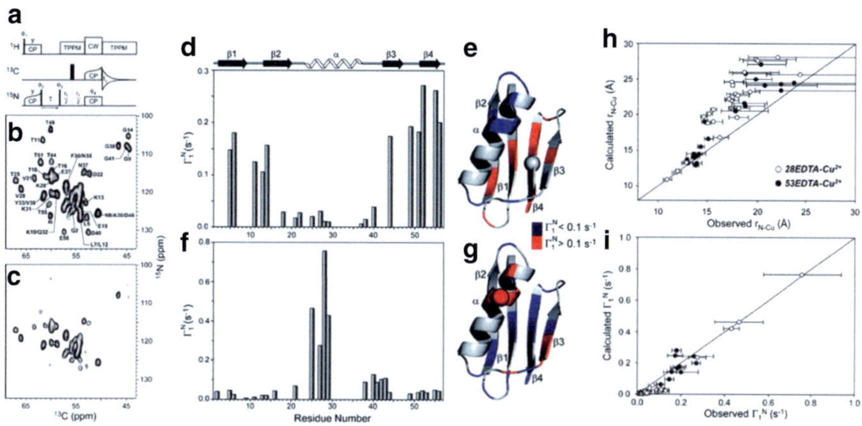

Fig. 27 (**a**) Solid-state NMR pulse scheme for site-resolved measurements of ^{15}N R_1 values. The experiment involves the acquisition of a series of 2D NCA correlation spectra, as a function of ^{15}N longitudinal relaxation delay τ. (**b**) Reference ($\tau = 0$) 2D NCA spectrum of 53EDTA-Cu^{II} with selected assignments indicated. (**c**) 2D NCA spectrum of 53EDTA-Cu^{II} with $\tau = 4$ s. (**d–g**) Backbone amide ^{15}N longitudinal paramagnetic relaxation enhancements, R_1^N, for 53EDTA-Cu^{II} (**d**, **e**) and for 28EDTA-Cu^{II} (**f**, **g**) are plotted as a function of residue number (**d–f**) and mapped onto the structure of GB1(**e–g**). (**h**, **i**) Comparison between the experimentally observed (**h**) $r_{N\text{–}Cu}$ and (**i**) R_1^N values, and the corresponding values derived from structural models of 28EDTA-M and 53EDTA-M proteins. Reprinted with permission from [101], © 2009 American Chemical Society

with Mn^{II}, which interacts with the anionic phospholipid head groups, the distances between different sites on the membrane-embedded peptide and the membrane surface were determined, along with the result that the membrane thickness is reduced in the proximity of an inserted molecule.

Finally, paramagnetic relaxation was used in order to determine the effects of metal ion association with amyloid fibrils. Several specific sites of Cu^{II} were detected in the amyloid protein Aβ(1–40) using PRE and shift perturbation data from solid-state NMR, and corroborated using detailed molecular dynamics (MD) models. It was also found by this means that the amyloid fibril structure is not significantly altered by Cu^{II} binding. This study is particularly interesting on account of its medical pertinence, as Cu^{II} is often found at elevated concentration in cells containing Alzheimers plaques [103].

4.4 Paramagnetic Centers and NMR Signal Enhancement: Dynamic Nuclear Polarization

Interest in paramagnetism in NMR has recently experienced an additional impulse due to the advent of high-field dynamic nuclear polarization (DNP) methods. The sensitivity of NMR poses a major barrier to the structural characterization of challenging active species. DNP provides a powerful way to deal with sensitivity issues. DNP allows transfer of polarization from the unpaired electrons of a paramagnetic center to the surrounding nuclei, and can enhance the sensitivity of NMR experiments by several orders of magnitude. Although the principles of this effect have been known for many decades [104–106], the development of its transposition to modern high-resolution NMR in liquids or solids only started to become operational a few years ago, first for liquids and now for solids. DNP at high fields requires relatively powerful and stable microwave sources, like gyrotrons (to polarize the electrons), and requires utilizing temperatures that are low enough so that the relative position of the electronic and nuclear spins are frozen. Commercial systems for liquids and solids have recently become available (in solids, for example, combining low-temperature MAS with in situ microwave irradiation) [107], opening up completely new application areas of investigation, previously forbidden for sensitivity reasons. Because it provides such a dramatic sensitivity enhancement, solid-state DNP NMR is currently emerging as a unique tool to study samples that are available in limited amounts or low concentrations and previously inaccessible to NMR studies [108]. Impressive DNP enhancements have been observed in biological samples such as amyloid fibrils [109] or membrane proteins [110, 111], as well as in materials sciences [112, 113]. These impressive gains clearly herald DNP as becoming a widespread technique in the next several years [108].

5 Concluding Remarks

The characterization of metal ions and their environments is one of the greatest challenges of modern chemistry and biology. In this chapter we have seen how recent progress in solid-state NMR has removed most of the barriers to progress in the structural characterization at the atomic level of samples containing paramagnetic metal ions. This has enabled previously inconceivable studies of new classes of molecules of high chemical and biological relevance containing paramagnetic metal ions, such as organometallic complexes, metalloproteins, batteries, and magnetic materials.

Acknowledgments A big “thanks” to Lyndon Emsley, Anne Lesage, Moreno Lelli, Andrew J. Pell, and Michael J. Knight, as well as to Ivano Bertini, Claudio Luchinat, Roberta Pierattelli, and Isabella C. Felli. The work on paramagnetic NMR in Lyon has been supported by the Agence Nationale de la Recherche (ANR 08-BLAN-0035-01 and 10-BLAN-0713-1) and by Joint Research Activity in the 6th and 7th Framework Program of the EC (EU-NMR and Bio-NMR).

Appendix A. More Theory

The dipolar Hamiltonian of (19) becomes

$$\tilde{\mathcal{H}}^{\mathrm{D}} = \left\{\frac{2}{3}\mathcal{D}_{2,0}\,\chi_{2,0} - \frac{1}{2}(\mathcal{D}_{2,1}\chi_{2,-1} + \mathcal{D}_{2,-1}\chi_{2,1}) - \frac{\sqrt{2}}{3}\mathcal{D}_{2,0}\,\chi_{0,0}\right\} I_z B_0. \tag{30}$$

In order to evaluate the transformation properties of the hyperfine Hamiltonian, this sum of products needs to be further simplified. This can be achieved by expressing the pair-wise products of the components $\mathcal{D}_{2,m}\,\chi_{2,-m}$ of two tensors of rank 2 in terms of the components of a product tensor $W_{l,0}$ of rank $l = 0 - 4$ by using the Clebsch–Gordan expansion [114, 115]:

$$\mathcal{D}_{2,\mathrm{m}}\,\chi_{2,-\mathrm{m}} = \sum_{l=0}^{4}\begin{pmatrix} 2 & 2 & l \\ -m & m & 0 \end{pmatrix} W_{l,0}, \tag{31}$$

where the term in parentheses represents the 3j symbols. Similarly, for the last term of (30):

$$\mathcal{D}_{2,0}\,\chi_{0,0} = \begin{pmatrix} 0 & 2 & 2 \\ 0 & 0 & 0 \end{pmatrix} W_{2,0}. \tag{32}$$

When applied to the Hamiltonian of (31), the antisymmetric terms of the expansion (l = 1, 3) do not need to be considered. Therefore, the above formulas give

$$\begin{aligned}\tilde{\mathcal{H}}^{\mathrm{D}} &= \left\{\frac{2}{3}\left(\frac{1}{\sqrt{5}}W_{0,0} - \sqrt{\frac{2}{7}}W_{2,0} + \sqrt{\frac{18}{35}}W_{4,0}\right) + \left(\frac{1}{\sqrt{5}}W_{0,0} - \frac{1}{\sqrt{14}}W_{2,0} - 2\sqrt{\frac{2}{35}}W_{4,0}\right)\right.\\ &\left.\quad - \frac{\sqrt{2}}{3}W_{2,0}\right\} I_z B_0 = \left\{\frac{\sqrt{5}}{3}W_{0,0} - \sqrt{\frac{7}{18}}W_{2,0}^{(\Delta\chi)} - \frac{\sqrt{2}}{3}W_{2,0}^{(\chi)}\right\} I_z B_0\\ &= -\left(\delta^{\mathrm{PC}} + \sqrt{\frac{2}{3}}\Delta\sigma^{(\Delta\chi)}d_{0,0}^{2}(\beta_{\mathrm{XL}}) + \sqrt{\frac{2}{3}}\Delta\sigma^{(\chi)}d_{0,0}^{2}(\beta_{\mathrm{DL}})\right)\tilde{T}_{1,0},\end{aligned} \tag{33}$$

where the superscripts χ and $\Delta\chi$ denote the contributions to the anisotropy that originate from the isotropic and anisotropic parts of the X tensor, respectively. It is interesting to note that the final form of the hyperfine interaction is again in all respects analogous to that of a diamagnetic chemical shift. Notably, the $W_{4,0}$ term vanishes, in contrast to, for example, the second-order quadrupolar effect where the product of two second-rank components contributes to the $W_{4,0}$. The rank-zero interaction is the isotropic shift and the rank-two components represent a shift anisotropy analogous to the diamagnetic CSA (the "dipolar shift anisotropy" or DSA).

The rank-zero portion of the dipolar interaction is the PCS δ^{PC}. Its full expression can be obtained expanding back the $W_{0,0}$ term as a product of two second-rank tensors according to the inverse of (31):

$$\begin{aligned} W_{0,0} &= \sum_{m=-2}^{2} \begin{pmatrix} 2 & 2 & 0 \\ m & -m & 0 \end{pmatrix} \mathcal{D}_{2,m}\, \chi_{2,-m} \\ &= \sum_{m=-2}^{2} \frac{(-1)^m}{\sqrt{5}} \mathcal{D}_{2,m}\, \chi_{2,-m}. \end{aligned} \tag{34}$$

After including the transformation Ω_{DX} to move from the PAS of the D tensor to that of the X tensor, we get

$$W_{0,0} = \frac{1}{\sqrt{5}} \mathcal{D}_{2,0} \sum_{m'=-2}^{2} \mathcal{D}^2_{0m'} \chi_{2,m'} \tag{35}$$

so that we obtain

$$\begin{aligned} \delta^{\mathrm{PC}} &= \frac{1}{3\hbar\gamma_{\mathrm{I}}} \mathcal{D}_{2,0} \sum_{m'=-2}^{2} \mathcal{D}^2_{0m'} \chi_{2,m'} \\ &= \frac{\sqrt{6}}{12\pi r^3} \left[\left(\frac{3\cos^2\theta - 1}{2} \sqrt{\frac{2}{3}} \Delta\chi_{\mathrm{ax}} \right) + \left(\sqrt{\frac{3}{8}} \sin^2\theta\, \mathrm{e}^{-\mathrm{i}2\varphi} \frac{1}{2} \Delta\chi_{\mathrm{rh}} \right) + \left(\sqrt{\frac{3}{8}} \sin^2\theta\, \mathrm{e}^{\mathrm{i}2\varphi} \frac{1}{2} \Delta\chi_{\mathrm{rh}} \right) \right] \cdot \\ &= \frac{1}{12\pi r^3} \left\{ \Delta\chi_{\mathrm{ax}} (3\cos^2\theta - 1) + \frac{3}{2} \Delta\chi_{\mathrm{rh}} \sin^2\theta \cos 2\varphi \right\} \end{aligned} \tag{36}$$

The first component of the shift anisotropy is

$$\Delta\sigma^{(\chi)} = \frac{1}{\mu_0 \mu_{\mathrm{B}} g_{\mathrm{e}}} \frac{\sqrt{2}}{3\hbar\gamma_{\mathrm{I}}} \mathcal{D}_{2,0} \chi_{0,0}, \tag{37}$$

which, in the PAS of the dipolar interaction, becomes

$$\Delta\sigma^{(\chi)} = \frac{1}{2\pi \mu_0 \mu_{\mathrm{B}} g_{\mathrm{e}}} \frac{\chi_{\mathrm{iso}}}{r^3}. \tag{38}$$

References

1. Mehring M (1983) Principles of high-resolution NMR in solids. Springer, Berlin
2. Schmidt-Rohr K, Spiess H (1994) Multidimensional solid-state NMR and polymers. Academic, London
3. Tycko R, Grey C (2009) Solid-state NMR in biological and materials physics. Phys Today 62:44–49
4. Blanc F, Coperet C, Lesage A, Emsley L (2008) High resolution solid state NMR spectroscopy in surface organometallic chemistry: access to molecular understanding of active sites of well-defined heterogeneous catalysts. Chem Soc Rev 37:518–526
5. McDermott AE (2009) Structure and dynamics of membrane proteins by magic angle spinning solid-state NMR. Annu Rev Biophys 38:385–403
6. Bertini I, Luchinat C, Parigi G (2001) Solution NMR of paramagnetic molecules; application to metallobiomolecules and models. Elsevier, London
7. Köhler FH (2001) Probing spin density by using NMR spectroscopy. In: Magnetism: molecules to materials. Wiley VCH, New York, pp 379–430
8. Bertini I, Luchinat C, Parigi G (2002) Paramagnetic constraints: an aid for quick solution structure determination of paramagnetic metalloproteins. Concepts Magn Reson 14:259–286
9. Otting G (2010) Protein NMR using paramagnetic ions. Annu Rev Biophys 39:387–405
10. Slichter C (1996) Principles of magnetic resonance. Springer, Berlin
11. Ganapathy S, Chacko V, Bryant R, Etter M (1986) Carbon CP-MASS NMR and X-ray crystal structure of paramagnetic lanthanide acetates. J Am Chem Soc 108:3159–3165
12. Haw JF, Campbell GC (1986) Temperature-dependent chemical shifts in ^{13}C CP/MAS spectra of paramagnetic solids. J Magn Reson 66:558–561
13. Campbell GC, Crosby RC, Haw JF (1986) ^{13}C chemical shifts which obey the Curie law in CP/MAS NMR spectra. The first CP/MAS NMR chemical-shift thermometer. J Magn Reson 69:191–195
14. Cheetham AK, Dobson CM, Grey CP, Jakeman RJB (1987) Paramagnetic shift probes in high-resolution solid-state NMR. Nature 328:706–707
15. Grey CP, Dobson CM, Cheetham AK, Jakeman RJB (1989) Studies of rare-earth stannates by ^{1}19Sn MAS NMR. The use of paramagnetic shift probes in the solid-state. J Am Chem Soc 111:505–511
16. Grey CP, Smith ME, Cheetham AK, Dobson CM, Dupree R (1990) ^{8}9Y MAS NMR study of rare-earth pyrochores: paramagnetic shifts in the solid state. J Am Chem Soc 112:4670–4675
17. Moon S, Patchkovskii S (2004) First-principles calculations of paramagnetic NMR shifts. In: Kaupp M, Bühl M, Malkin VG (eds) Calculation of NMR and EPR parameters. Theory and applications. Wiley-VCH, Mannheim, pp 325–338
18. Hrobarik P, Reviakine R, Arbuznikov AV, Malkina OL, Malkin VG, Köhler FH, Kaupp M (2007) Density functional calculations of NMR shielding tensors for paramagnetic systems with arbitrary spin multiplicity: validation on 3d metallocenes. J Chem Phys 126:024107
19. Repisky M, Komorovsky S, Malkin E, Malkina OL, Malkin VG (2010) Effects of finite size nuclei in relativistic four-component calculations of hyperfine structure. Chem Phys Lett 488:94
20. Pennanen TO, Vaara J (2008) Nuclear magnetic resonance chemical shift in an arbitrary electronic spin state. Chem Phys Lett 100:133002
21. Samoson A, Tuherm T, Gan Z (2001) High-field high-speed MAS resolution enhancement in H1 NMR spectroscopy of solids. Solid State Nucl Magn Reson 20:130–136
22. Alla M, Lippmaa E (1982) Resolution limits in magic-angle rotation NMR-spectra of polycrystalline solids. Chem Phys Lett 87:30–33
23. Brough A, Grey C, Dobson C (1993) Paramagnetic ions as structural probes in solid-state NMR: distance measurements in crystalline lanthnide acetates. J Am Chem Soc 115:7318–7327

24. Solomon I (1955) Relaxation processes in a system of 2 spins. Phys Rev 99:559–565
25. Gueron M (1975) Nuclear-relaxation in macromolecules by paramagnetic-ions – novel mechanism. J Magn Reson 19:58–66
26. Kervern G, Steuernagel S, Engelke F, Pintacuda G, Emsley L (2007) Absence of curie relaxation in paramagnetic solids yields long H1 coherence lifetimes. J Am Chem Soc 129:14118–14119
27. Kervern G, Pintacuda G, Zhang Y, Oldfield E, Roukoss C, Kuntz E, Herdtweck E, Basset JM, Cadars S, Emsley L (2006) Solid-state NMR of a paramagnetic DIAD-Fe-II catalyst: sensitivity, resolution enhancement, and structure-based assignments. J Am Chem Soc 128:13545–13552
28. Clayton AN, Dobson CM, Grey CP (1990) High-resolution C-13 MAS NMR spectra of paramagnetic lanthanide complexes. J Chem Soc Chem Comm 72–74
29. Liu K, Ryan D, Nakanishi K, McDermott A (1995) Solid-state NMR-studies of paramagnetic coordination-complexes – a comparison of protons and deuterons in detection and decoupling. J Am Chem Soc 117:6897–6906
30. Ishii Y, Wickramasinghe NP, Chimon S (2003) A new approach in 1D and 2D C13 high-resolution solid-state NMR spectroscopy of paramagnetic organometallic complexes by very fast magic-angle spinning. J Am Chem Soc 125:3438–3439
31. Wickramasinghe NP, Shaibat M, Ishii Y (2005) Enhanced sensitivity and resolution in H1 solid-state NMR spectroscopy of paramagnetic complexes under very fast magic angle spinning. J Am Chem Soc 127:5796–5797
32. Wickramasinghe NP, Ishii Y (2006) Sensitivity enhancement, assignement and distance measurement in 13C solid-state NMR spectroscopy for paramagnetic systems under fast magic angle spinning. J Magn Reson 181:233–243
33. Laage S, Sachleben JR, Steuernagel S, Pierattelli R, Pintacuda G, Emsley L (2009) Fast acquisition of multidimensional NMR experiments enabled by ultra-fast MAS. J Magn Reson 196:133–141
34. Laage S, Marchetti A, Sein J, Pierattelli R, Sass HJ, Grzesiek S, Lesage A, Pintacuda G, Emsley L (2008) Band-selective ^{1}H-^{13}C cross-polarization in ultra-fast biomolecular MAS solid-state NMR. J Am Chem Soc 130:17216–17217
35. Verel R, Ernst M, Meier BH (2001) Adiabatic dipolar recoupling in solid-state NMR: the DREAM scheme. J Magn Reson 150:81–99
36. Bertini I, Emsley L, Lelli M, Luchinat C, Mao J, Pintacuda G (2010) Ultra-fast MAS solid-state NMR permits extensive ^{13}C and ^{1}H detection in paramagnetic metalloproteins. J Am Chem Soc 132:5558–5559
37. Wickramasinghe NP, Shaibat MA, Jones CR, Casabianca LB, de Dios AC, Harwood JS, Ishii Y (2008) Progress in C13 and H1 solid-state nuclear magnetic resonance for paramagnetic systems under very fast magic angle spinning. J Chem Phys 128:52210
38. Garwood M, DelaBarre L (2001) The return of the frequency sweep: designing adiabatic pulses for contemporary NMR. J Magn Reson 153:155–177
39. Kervern G, Pintacuda G, Emsey L (2007) Fast adiabatic pulses for solid-state NMR of paramagnetic systems. Chem Phys Lett 435:157–162
40. Pell AJ, Kervern G, Emsley L, Deschamps M, Massiot D, Grandinetti PJ, Pintacuda G (2011) Broadband inversion for MAS NMR with single-sideband-selective adiabatic pulses. J Chem Phys 134:024117
41. Siegel R, Nakashima TT, Wasylishen RE (2007) Sensitivity enhancement of NMR spectra of half-integer spin quadrupolar nuclei in solids using hyperbolic secant pulses. J Magn Reson 184:85–100
42. Nakashima TT, Wasylishen RE, Siegel R, Ooms KJ (2008) Sensitivity enhancement of solid-state NMR spectra of half-integer spin quadrupolar nuclei: double- or single-frequency sweeps? Insights from the hyperbolic secant experiment. Chem Phys Lett 450:417–421
43. Nakashima TT, Teymoori R, Wasylishen RE (2009) Using hyperbolic secant pulses to assist characterization of chemical shift tensors for half-integer spin quadrupolar nuclei in MAS powder samples. Magn Reson Chem 47:465–471

44. Dey KK, Prasad S, Ash JT, Deschamps M, Grandinetti PJ (2007) Spectral editing in solid-state MAS NMR of quadrupolar nuclei using selective satellite inversion. J Magn Reson 185:326–330
45. Kervern G, D'Aleo A, Maury O, Emsley L, Pintacuda G (2009) Crystal structure determination of powdered paramagnetic lanthanide complexes by proton NMR. Angew Chem Int Ed Engl 48:3082–3086
46. Coronado E, Delhaes P, Gatteschi D, Miller JS (eds) (1996) Molecular magnetism: from molecular sssemblies to the devices. Kluwer, Dordrecht
47. Campbell GC, Haw JF (1988) Determination of magnetic and structural properties in solids containing antiferromagnetically coupled metal centers using NMR methods. magneto-structural correlations in anhydrous copper(II) n-butyrate. Inorg Chem 27:3706–3709
48. Campbell GC, Reibenspies JH, Haw JF (1991) Solid-state NMR studies of magneto-structural correlations in anhydrous copper(II) carboxylates. Inorg Chem 30:171–176
49. Köhler FH, Lescouëzec R (2004) Highly resolved spin-density distribution in the Prussian-blue precursors $Cs_2K[Fe(CN)_6]$ and $Cs_2K[Mn(CN)_6]$. Angew Chem Int Ed Engl 43:2571–2573
50. Köhler FH, Lescouëzec R (2009) Revisiting prussian blue analogues with MAS NMR spectroscopy: spin density and local structure in $Cd_3[Fe(CN)_6]$ $I5H_2O$. Angew Chem Int Ed Engl 48:1673–1676
51. Heise H, Köhler FH, Mota F, Novoa JJ, Veciana J (1999) Determination of the spin distribution in nitronylnitroxides by solid-state H-1, H-2, and C-13 NMR spectroscopy. J Am Chem Soc 121:9659–9667
52. Sporer C, Heise H, Wurst K, Ruiz-Molina D, Kopacka H, Jaitner P, Köhler F, Novoa JJ, Veciana J (2004) Magneto-structural characterization of metallocene-bridged nitronyl nitroxide diradicals by X-ray, magnetic measurements, solid-state NMR spectroscopy, and ab initio calculations. Chem Eur J 10:1355–1365
53. Ziessel R, Stroh C, Heise H, Köhler FH, Turek P, Claiser N, Souhassou M, Lecomte C (2004) Strong exchange interactions between two radicals attached to nonaromatic spacers deduced from magnetic, EPR, NMR, and electron density measurements. J Am Chem Soc 126:12604–12613
54. Mao JH, Zhang Y, Oldfield E (2002) Nuclear magnetic resonance shifts in paramagnetic metalloporphyrins and metalloproteins. J Am Chem Soc 124:13911–13920
55. Zhang Y, Sun HH, Oldfield E (2005) Solid-state NMR fermi contact and dipolar shifts in organometallic complexes and metalloporphyrins. J Am Chem Soc 127:3652–3653
56. Crozet M, Chaussade M, Bardet M, Emsley L, Lamotte B, Mouesca JM (2000) Carbon-13 solid-state NMR studies on synthetic model compounds of [4Fe-4S] clusters in the 2(+) state. J Phys Chem A 104:9990–10000
57. Rancurel C, Heise H, Köhler FH, Schatzschneider U, Rentschler E, Vidal-Gancedo J, Veciana J, Sutter JP (2004) Spin transfer and magnetic interaction via phosphorus in nitronyl nitroxide radical-substituted triphenylphosphine derivatives. J Phys Chem A 108:5903–5914
58. Grey CP, Dupré N (2004) NMR studies of cathode materials for lithium-ion rechargeable batteries. Chem Rev 104;4493–4512
59. Kim J, Middlemiss DS, Chernova NA, Zhu BYX, Masquelier C, Grey CP (2010) Linking local andvironments and hyperfine shifts: a combined experimental and theoretical ^{31}P and ^{7}Li solid-state NMR study of paramagnetic Fe(III) phosphates. J Am Chem Soc 132:16825–16840
60. Grey CP, Lee YJ (2003) Lithium MAS NMR studies of cathode materials for lithium-ion batteries. Solid State Sci 5:883–894
61. Koder RL, Walsh JD, Pometum MS, Dutton PL, Wittebort RJ, Miller AF (2006) ^{15}N solid-state NMR provides a sensitive probe of oxidized flavin reactive sites. J Am Chem Soc 128:15200–15208
62. Blümel J, Herker M, Hiller W, Köhler FH (1996) Study of paramagnetic chromocenes by solid-state NMR spectroscopy. Organometallics 15:3474–3476

63. Xing Q, Milius W, Krauss HL, Blümel J, Hilbig H, Köhler FH, Strauss W, Bayreuther G (1999) Darstellung und Eigenschaften von Chromorganylen aus Phillips-Katalysatoren und Ethylen. Z Anorg Allg Chem 625:521–529
64. Schnellbach M, Köhler FH, Blümel J (1996) The union carbide catalyst ($Cp_2Cr + SiO_2$), studied by solid-state NMR. J Organomet Chem 250:227–230
65. Estephane J, Groppo E, Vitillo JG, Damin A, Gianolio D, Lamberti C, Bordiga S, Quadrelli EA, Basset JM, Kervern G, Emsley L, Pintacuda G, Zecchina A (2010) A multi-technique approach to spin-flips for cp_2Cr(II) chemistry in confined state. J Phys Chem C 114:4451–4458
66. Shaibat MA, Casabianca LB, Wickramasinghe NP, Guggenheim S, de Dios AC, Ishii Y (2007) Characterization of polymorphs and solid-state reactions for paramagnetic systems by ^{13}C solid-state NMR and ab initio calculations. J Am Chem Soc 129:10968–10969
67. Bleaney B (1972) Nuclear magnetic-resonance shifts in solution due to lanthanide ions. J Magn Reson 8:91
68. Golding RM, Halton MP (1972) Paramagnetic lanthanide induced shifts. Aust J Chem 25:2577–2581
69. Jovanovic T, Farid R, Friesner RA, McDermott AE (2005) Thermal equilibrium of high- and low-spin forms of cytochrome P450BM-3: repositioning of the substrate? J Am Chem Soc 127:13548–13552
70. Jovanovic T, McDermott AE (2005) Observation of ligand binding to cytochrome P450 BM-3 by means of solid-state NMR spectroscopy. J Am Chem Soc 127:13816–13821
71. Kubo A, Spaniol TP, Terao T (1998) The effect of bulk magnetic susceptibility on solid state NMR spectra of paramagnetic compounds. J Magn Reson 133:330–340
72. Spaniol TP, Kubo A, Terao T (1999) Resolution enhancement of magic-angle spinning NMR spectra for paramagnetic solids by zero-quantum NMR. Mol Phys 96:827–834
73. Balayssac S, Bertini I, Lelli M, Luchinat C, Maletta M (2007) Paramagnetic ions provide structural restraints in solid-state NMR of proteins. J Am Chem Soc 129:2218–2219
74. Balayssac S, Bertini I, Falber K, Fragai M, Jehle S, Lelli M, Luchinat C, Oschkinat H, Yeo KJ (2007) Solid-state NMR of matrix metalloproteinase 12: an approach complementary to solution NMR. Chembiochem 8:486–489
75. Balayssac S, Bertini I, Bhaumik A, Lelli M, Luchinat C (2008) Paramagnetic shifts in solid-state NMR of proteins to elicit structural information. Proc Natl Acad Sci USA 45:17284–17289
76. Bloembergen N (1950) Fine structure of the magnetic resonance line of protons in $CuSO_4$ $5H_2O$. Physica 16:95–112
77. Rundle RE (2002) Antiferromagnetic ordering and electronic structure of $CuCl_2$-$2H_2O$ as determined from nuclear magnetic resonancel. J Am Chem Soc 79:3372–3374
78. Poulis NJ, Hardeman GEG (1952) Physica 18:201
79. McGarvey BR, Nagy S (1987) Evidence for an anisotropic contact shift, proton NMR study of line shapes in uranocenes and $(C_5H_5)_3$ UCl powders. Inorg Chem 26:4198–4203
80. McGarvey BR, Nagy S (1987) Anisotropy in the proton NMR paramagnetic shift of cyclooctatetraene and cyclopentadiene anions coordinated to U(IV). Inorg Chim Acta 139:319–321
81. Nayeem A, Yesinowski JP (1988) Calculation of magic-angle spinning nuclear magnetic-resonance spectra of paramagnetic solids. J Chem Phys 89:4600–4608
82. Brough AR, Grey CP, Dobson CM (1992) Structural information from NMR-studies of paramagnetic solids – Na-23 MAS spectra of sodium lanthanide salts of ethylenediaminete-traacetic acid. J Chem Soc Chem Comm 742–743
83. Lee YJ, Grey CP (2002) Determining the lithium local environments in the lithium manganates $LiZn_{0.5}Mn_{1.5}O_4$ and Li_2MnO_3 by analysis of the ^6Li MAS NMR spinning sideband manifolds. J Phys Chem B 106:3576–3582
84. Huang WL, Schopfer M, Zhang C, Howell RC, Gee BA, Francesconi LC, Polenova T (2006) Probing local environments in paramagnetic Europium-substituted Keggin solids by P-31 magic angle spinning NMR spectroscopy. J Phys Chem B 110:12340–12350

85. Huang W, Schopfer M, Zhang C, Howell R, Todaro L, Gee B, Francesconi L, Polenova T (2008) ^{31}P magic angle spinning NMR spectroscopy of paramagnetic rare-earth-substituted Keggin and Wells–Dawson solids. J Am Chem Soc 130:481–490
86. Lin TH, Dinatale JA, Void RR (1994) Determination of molecular-reorientation rates and electron-nuclear coupling in paramagnetic materials by deuterium solid echo NMR-spectroscopy. J Am Chem Soc 116:2133–2134
87. Liu K, Williams J, Lee HR, Fitzgerald MM, Jensen GM, Goodin DB, McDermott AE (1998) Solid-state deuterium NMR of imidazole ligands in cytochrome c peroxidase. J Am Chem Soc 120:10199–10202
88. Spaniol TP, Kubo A, Terao T (1997) Two-dimensional deuterium magic-angle-spinning nuclear magnetic resonance of paramagnetic compounds: separation of para-magnetic and quadrupole interactions. J Chem Phys 106:5393–5405
89. Antonijevic S, Wimperis S (2005) Separation of quadrupolar and chemical/paramagnetic shift interactions in two-dimensional ^{2}H ($I = 1$) nuclear magnetic resonance spectroscopy. J Chem Phys 122:044312
90. Lee H, Polenova T, Beer R, McDermott A (1999) Line-shape fitting of deuterium magic angle spinning spectra of paramagnetic compounds in slow and fast limit motion regimes. J Am Chem Soc 121:6884–6894
91. Lee H, Ortiz de Montellano PR, McDermott A (1999) Deuterium magic angle spinning studies of substrates bound to cytochrome P450. Biochemistry 38:10808–10813
92. Wickramasinghe NP, Shaibat MA, Ishii Y (2007) Elucidating connectivity and metal-binding structures of unlabeled paramagnetic complexes by ^{13}C and ^{1}H solid-state NMR under fast magic angle spinning. J Phys Chem B 111:9693–9696
93. Heise H, Köhler FH, Xie XL (2001) Solid-state NMR spectroscopy of paramagnetic metallocenes. J Magn Reson 150:198–206
94. Heise H, Köhler FH, Herker M, Hiller W (2002) Inter-and intramolecular spin transfer in molecular magnetic materials. solid-state NMR spectroscopy of paramagnetic metallocenium ions. J Am Chem Soc 124:10823–10832
95. Dobson CM, Goberdhan DGC, Ramsey JDF, Rodger SA (1988) ^{29}Si MAS NMR study of the hydration of tricalcium silicate in the presence of finely divided silica. J Mater Sci 23:4108–4114
96. Vijayan V, Demers JP, Biernat J, Mandelkow E, Becker S, Lange A (2009) Low-power solid-state NMR experiments for resonance assignment under magic-angle spinning. Chemphyschem 10:2205–2208
97. Wickramasinghe NP, Kotecha M, Samoson A, Past J, Ishii Y (2007) Sensitivity enhancement in ^{13}c solid-state NMR of protein microcrystals by use of paramagnetic metal ions for optimizing ^{1}H T_1 relaxation. J Magn Reson 184:350–356
98. Wickramasinghe NP, Parthasarathy S, Jones CR, Bhardwaj C, Long F, Kotecha M, Mehboob S, Fung LWM, Past J, Samoson A, Ishii Y (2009) Nanomole-scale protein solid-state NMR by breaking intrinsic $^{1}HT_1$ boundaries. Nat Meth 6:215–218
99. Nadaud PS, Helmus JJ, Sengupta I, Jaroniec CP (2010) Rapid acquisition of multidimensional solid-state NMR spectra of proteins facilitated by covalently bound paramagnetic tags. J Am Chem Soc 132:9561–9563
100. Nadaud PS, Helmus JJ, Hofer N, Jaroniec CP (2007) Long-range structural restraints in spin-labeled proteins probed by solid-state nuclear magnetic resonance spectroscopy. J Am Chem Soc 129:7502
101. Nadaud PS, Helmus JJ, Kall SL, Jaroniec CP (2009) Paramagnetic ions enable tuning of nuclear relaxation rates and provide long-range structural restraints in solid-state NMR of proteins. J Am Chem Soc 131:8108–8120
102. Buffy JJ, Hong T, Yamaguchi S, Waring AJ, Lehrer RI, Hong M (2003) Solid-state NMR investigation of the depth of insertion of protegrin-1 in lipid bilayers using paramagnetic Mn^{2+}. Biophys J 85:2363–2373

103. Parthasarathy S, Long F, Miller Y, Xiao Y, McElheny D, Thurber K, Ma B, Nussinov R, Ishii Y (2011) Molecular-level examination of Cu^2+ binding structure for amyloid fibrils of 40-residue Alzheimer's β by solid-state NMR spectroscopy. J Am Chem Soc 133:3390–3400
104. Carver TR, Slichter CP (1953) Polarization of nuclear spins in metals. Phys Rev 92:212–213
105. Carver TR, Slichter CP (1956) Experimental verification of the Overhauser nuclear polarization effect. Phys Rev 102:975–980
106. Slichter CP (2010) The discovery and demonstration of dynamic nuclear polarization – a personal and historical account. Phys Chem Chem Phys 12:5741–5751
107. Rosay M, Tometich L, Pawsey S, Bader R, Schauwecker R, Blank M, Borchard PM, Cauffman SR, Felch KL, Weber RT, Temkin RJ, Griffin RG, Maas WE (2010) Solid-state dynamic nuclear polarization at 263 Ghz: spectrometer design and experimental results. Phys Chem Chem Phys 12:5850–5855
108. Griffin GR, Prisner TF (2010) High field dynamic nuclear polarization – the renaissance. Phys Chem Chem Phys 12:5737–5743
109. Debelouchina GT, Bayro MJ, van der Wel PCA, Caporini MA, Barnes AB, Rosay M, Maas WE, Griffin RG (2010) Dynamic nuclear polarization-enhanced solid-state NMR spectroscopy of GNNQQNY nanocrystals and amyloid fibrils. Phys Chem Chem Phys 12:5911–5919
110. Bajaj VS, Mak-Jurkauskas ML, Belenky M, Herzfeld J, Griffin RG (2010) DNP enhanced frequency-selective TEDOR experiments in bacteriorhodopsin. J Magn Reson 202:9–13
111. Salnikov E, Rosay M, Pawsey S, Ouari O, Tordo P, Bechinger B (2010) Solid-state NMR spectroscopy of oriented membranepolypeptides at 100 K with signal enhancement by dynamic nuclear polarization. J Am Chem Soc 132:5940–5941
112. Lesage A, Lelli M, Gajan D, Caporini MA, Vitzthum V, Mieville P, Alauzun J, Roussey A, Thieuleux C, Mehdi A, Bodenhausen G, Coperet C, Emsley L (2010) Surface enhanced NMR spectroscopy by dynamic nuclear polarization. J Am Chem Soc 132:15459–15461
113. Lelli M, David G, Lesage A, Caporini M, Vitzthum V, Mieville P, Heroguel F, Rascon F, Roussey A, Thieuleux C, Boualleg M, Veyre L, Bodenhausen G, Coperet C, Emsley L (2011) Fast characterization of functionalized silica materials by Silicon-29 surface-enhanced NMR spectroscopy using dynamic nuclear polarization. J Am Chem Soc 133:2104–2107
114. Brink D, Satchler G (1968) Angular momentum. Oxford University Press, London
115. Rose M (1967) Elementary theory of angular momentum. Wiley, New-York

Index

MIX
Papier aus verantwortungsvollen Quellen
Paper from responsible sources
FSC® C105338

If you have any concerns about our products,
you can contact us on
ProductSafety@springernature.com

In case Publisher is established outside the EU,
the EU authorized representative is:
Springer Nature Customer Service Center GmbH
Europaplatz 3, 69115 Heidelberg, Germany

Printed by Libri Plureos GmbH
in Hamburg, Germany